THE MELANCHOLY ASSEMBLAGE

THE MELANCHOLY ASSEMBLAGE

Affect and Epistemology in the English Renaissance

Drew Daniel

Fordham University Press
New York 2013

Library of Congress Cataloging-in-Publication Data

Daniel, Drew, 1971–
The melancholy assemblage : affect and epistemology in the English Renaissance / Drew Daniel.
pages cm
Summary: "This book considers melancholy as an "assemblage," as a network of dynamic, interpretive relationships between persons, bodies, texts, spaces, structures, and things. In doing so, it parts ways with past interpretations of melancholy. Tilting the English Renaissance against the present moment, Daniel argues that the basic disciplinary tension between medicine and philosophy persists within contemporary debates about emotional embodiment. To make this case, the book binds together the paintings of Nicholas Hilliard and Isaac Oliver, the drama of Shakespeare, the prose of Burton, and the poetry of Milton. Crossing borders and periods, Daniel combines recent theories which have—until now—been regarded as incongruous by their respective advocates. Asking fundamental questions about how the experience of emotion produces community, the book will be of interest to scholars of early modern literature, psychoanalysis, the affective turn, and continental philosophy"—Provided by publisher.
Includes bibliographical references and index.
ISBN 978-0-8232-5127-8 (hardback) — ISBN 978-0-8232-5128-5 (paper)
1. English literature—Early modern, 1500–1700—History and criticism—Theory, etc. 2. Affect (Psychology) in literature. 3. Knowledge, Theory of, in literature. 4. Literature and science—Great Britain—History. 5. Art and literature—Great Britain—History. 6. Science in literature. 7. Science—Philosophy. 8. Renaissance—England. I. Title.
PR421.A425D36 2013
820.9'353—dc23

2012043868

Printed in the United States of America

15 14 13 5 4 3 2 1

First edition

To my parents and stepparents

Contents

Illustrations

Acknowledgments

> You can't detach what you seek to know
> from your emotions as you seek to know it.
>
> —*Lyn Hejinian, "The Fatalist"*[1]

The first assemblage encountered in this book is distinctly sanguine in temperament—a modular network of students, colleagues, mentors, allies, editors, family members, band mates, and friends. Many institutions have hosted me while I assembled this work: the University of California at Berkeley; the Huntington Library; the Folger Library; St. John's College, Oxford; and Johns Hopkins University. There I have had the distinct pleasure of learning from feisty undergraduates and being regularly challenged by the participants in a monthly early modern reading group whose cohort includes Rebecca Buckham, Marisa O'Connor, Jacob Chilton, David Hershinow, Will Miller, Benjamin Parris, Andrew Sisson, and Maggie Vinter. My present and former colleagues in the Department of English have been extraordinary allies and famously demanding interlocutors: The book would not have taken the shape it did had I not had the scholarly examples of Amanda Anderson, Sharon Cameron, Simon During, Frances Ferguson, Jared Hickman, Douglas Mao, Christopher Nealon, Jesse Rosenthal, Eric Sundquist, and Mark Thompson before me. I am also deeply grateful to three colleagues at Hopkins from outside my department who have responded to early versions of this work with expertise, encouragement, and timely advice: Jane Bennett, Stephen Campbell, and Katrin Pahl.

Outside of my own institution, there stands a constellation of people who have encouraged, hosted, helped, or inspired in manifold ways: Richard

McCabe, Gerard Passannante, Steven Shaviro, Molly Murray, Jeffrey Cohen, Nigel Smith, Stephen Greenblatt, Eileen Joy, Fred and Katherine Maus, Scott Wilson, Tavia Nyong'o, Heather Love, Jose Muñoz, Lauren Berlant, Jonathan Flatley, Bryan Lowrance, Madhavi Menon, Jonathan Gil Harris, Gustavus Stadler, Homay King, Jonathan Dollimore, Monica Youn, Erika Clowes, Will Stockton, Christopher Pye, Steve Burt, Ken Wissoker, Mark Wunderlich, and Francois-Xavier Gleyzon have nothing in common with each other but my gratitude, dispensed in variably sized portions. Versions of this project have been given as talks at Columbia, Princeton, UCLA, UC Irvine, George Washington University, and the University of Virginia, and I am also grateful to my hosts and respondents at those institutions. Graham Hammill and Jacques Lezra wrote responses to the version of my chapter on "The Merchant of Venice," which was published in *Shakespeare Quarterly*, and I am doubly indebted for their critical and constructive engagement.

Whether reading chapters, hosting talks, curating conferences, or throwing dinner parties, Julia Reinhard Lupton and Kenneth Reinhard have been particularly staunch allies, collaborators, and friends: This project has been shaped by their example and their tough questions. I also wish to thank my manuscript's two readers for their encouragement, advice, and critiques, and to preemptively absolve them of any errors and lacunae that remain.

I want to thank the editors of *Shakespeare Quarterly* for permission to reprint my chapter on "The Merchant of Venice." I also wish to thank Margaret Gray of Powis Castle, Shannon Haskett at Patrick Painter Editions, and the helpful staff of Art Resource and the Bridgeman Art Library. An extra special thanks to the Tirkanits/Daniel Foundation for the Arts for making the color insert possible.

My greatest intellectual and emotional debts are to the two advisors to this project who stewarded me through the writing process and out the other side: Richard Halpern and Janet Adelman offered me paternal/maternal advice, ferocious criticism, and crucial support. From separate corners and opposite coasts, they each personified scholarship as a lifelong practice of creative commitment to the pleasure of difficulty. Janet's death leaves a wound twice over, within Shakespearean scholarship and psychoanalytic criticism, but the monument of her work holds fast.

Arriving at the other end of a seemingly endless process, I wish to thank my editors at Fordham University Press—Helen Tartar, Thomas Lay, Nicole Pfeifer, and Eric Newman—for believing in the project and steering it toward publication with flexibility, honesty, and good sense.

The enduring support of my family has been both inspiring and humbling. Over the past decade, Rollin Daniel, Beatrix Tirkanits, Kathleen Daniel, Marty Sussman, and Matthew Sussman have patiently endured more than the legal limit of melancholizing. I want to thank them for reminding me of joy.

It is not possible to thank Martin Schmidt for twenty years of unrelenting honesty and love. Instead, I will repeat what Jorge Luis Borges said of De Quincey—"my debt to him is so vast that to specify a part of it seems to repudiate or to silence the others."[2] To Martin I dedicate more than can be expressed in language.

THE MELANCHOLY ASSEMBLAGE

Introduction

Sweet peas in dark gardens
Squirt false melancholy over history.

—*John Ashbery, "Album Leaf"*[1]

False Squirts and Panic Attacks

The topic of melancholy courts a suitable exhaustion at the present moment. Faced with the prospect of a sequence of readings of early modern representations of this all-too-familiar emotional stance, one might well wonder whether anything could have gone unnoticed about this particular quintessence of scholarly dust. If, to take up the Ashbery poem's sharp formulation, the dark gardens of history are rendered faintly slapstick by the humoral stain of "false melancholy" squirted upon them, one might well ask: Given the now ascendant functionalist understanding of emotions as neurochemical events, was melancholy ever *not* false? What is to be done with the theoretical deadweight of the humors today? Why should these long discredited frameworks of Renaissance intellectual history detain anyone further?

Whether we understand the apparent "falseness" of melancholy as an aesthetic matter of passé sensibility or an epistemological matter of insincerity and unverifiability or a biological matter of bad science, the question of how to categorize such a culturally productive mistake has substantial critical and philosophical implications. The interpretation of melancholy turns upon a basic human question whose pressure persists within the present: What happens when we encounter the emotion of another? I want to pursue this via a narrative detour, which I shall use as both an allegory of affective mindreading and a phenomenological "intuition pump."

While walking down the hallways of the Modern Language Association convention in Philadelphia, I pass within earshot of a woman violently shouting into a cell phone. Ringing down the yellow corridors of the hotel and leaking into the conference rooms in which sessions on affect theory and disability studies are already underway, her voice shakes with naked emotional distress:

> I'm having a panic attack, right now, and I need to know if you can help me. Can you help me? Can you come here right now and help me stop this? I need to know if you can help me *right now*!

Reacting to the cry, I feel my breath stop and my heart rate speed up. My hands clench into sweaty fists as my eyes dart sideways, upward, and elsewhere. A tremor of sympathy and a reflex of disgust move through me as I momentarily consider and then reject the possibility that I stop and try to assist this distraught person. Policing my body to walk studiously away from the source of the alarming sound, I feel guilt at my own voracious interest in the raw need to be consoled contained within her words. I want to get closer, but I also feel an urge to hide or be elsewhere, to deny my eavesdropping on this suffering human being, as if acknowledging that experience of spectatorship all by itself could compound the painful circuit of self-exposure, shame, and vulnerability already underway. "Caring" in a way indistinguishable from "not caring," I assure myself that there is no need to invade her privacy and perhaps make things worse. I walk on.

Caught between sympathy and aversion but continuing to discipline my head and neck to face forward, it is clear to me that narcissism lurks behind my urge to look back at her: I need to know if she is simply a random hotel guest or another academic, and thus, crucially, someone "like me." This desire to look backward to see if she bears the convention badge indexes a basic need to constellate us into categories of likeness and distance. Shielding myself from identifying with her, I try to position her against the backdrop of a competitive profession that cultivates and mobilizes the productive affect of constant low-level anxiety. Is she a "distraught job seeker," a "publisher in a recession," a "tenure-obsessed junior faculty member," an "insecure senior scholar," or some other inhabitant in the conference's bestiary of depressed and anxious knowledge workers? Toying with these possible scenarios, I am struck by their cruel irrelevance to her human situation. The game of status and stratification distracts from more basic, unresolved questions: What had just happened, or was still happening, between us? What reserves of decorum and self-control does she owe me? What care and obligations do I owe her?

As a public expression of an inner emotional state, her voice occasions a complex network of performed speech-acts, embodied brain states, gestural reactions, affective flows, and philosophical dilemmas. Conscripted into the visual and acoustic radius of her emotional event, I am brought into some sort of loose perceptual community with her, a tenuous, infra-thin state of fellow feeling. But in evaluating those perceptions and their rightness of fit, I am just as inescapably pressed with a cascade of epistemological problems which I can neither resolve nor verify, but whose critical tenor tend to skeptically erode the conditions of that community: How much access do I really have to her emotional state? Does she need help? Who are we to each other? Is this a real crisis or not, and what sort of criteria can even resolve such a question?

The multiplicity of feelings and the questions of knowledge are both givens within the same situation. If there is a basic causal priority in which the event creates the perceptions and feelings that then trigger doubts about the validity or provenance of those perceptions and feelings, no conceptual priority of feeling over knowing follows from that causal chain. I perceive the emotion of another, and in that perception a delicate bond is formed between us. But in reflecting critically upon its source and status, I initiate an inquiry without any clear stopping point, and with no commitment to the necessity of preserving that bond. An affective bond-which-is-also-a-gap emerges. The outcome is indeterminate.

This book argues that melancholy constitutes a historically unique vector for a phenomenologically general problem encountered in the panic attack I witnessed. For Renaissance subjects no less than for us today, the intrinsically unverifiable nature of first person reports about emotional experience means that encounters with their display—even our own—can occasion reflection on the criteria we use to identify and assess the source of the display and the explanatory limits of those criteria.[2] When we encounter the emotion of another, we experience both an emergence of perceptual community and an epistemological problem of access to the emotional experience of others. It is not that stances loosely called "sympathy" and "skepticism" form two ends of a single polarity and we simply track our reactions along that arc. After all, we can completely recognize the emotion of another as a strongly felt emotion and yet feel no sympathy for it whatsoever (a drunken man in a rage over the result of a football match). Conversely, we can sense that some emotional expression is something of a put-on and yet nonetheless feel a deep sympathy for its sufferer regardless of the misleading surface (a child weeping and throwing a tantrum simply because she is tired).

Of course, not all emotions sue for their own validity: Unless we are afflicted by a particularly crippling form of paranoia or philosophical skepticism, we do not usually need to ask ourselves whether a given joyful or terrified person is really joyful or really terrified or simply pretending to be so. Yet the common and repellent experience of someone with a "fake laugh" offers a ready-to-hand example of the sort of epistemological tremor that shows up when an emotion arrives in a questionable shape. Since the notion of a "fake laugh" cannot get off the ground without a corresponding confidence about what would constitute a real laugh, we here brush up against what Michael Polanyi terms *the tacit dimension* of emotional recognition that underwrites our responses to borderline cases. When it comes to what makes a laugh fall on either side of this boundary, "we can know more than we can tell."[3] "Melancholy," no less than "panic attack," presupposes such borderline cases of failure, falsification, and doubt precisely because these words have also succeeded in making the emotions they supposedly designate intelligible to one's self and to others.

Melancholy is one component of the relational system of the humors, the ancient medical doctrine of four liquid substances flowing and mingling within the body. Because the subsequent history of science has stranded the scheme of the humors with phlogiston and ether in the purgatory of epistemic defeat, the distinction between genuine and pretended examples of melancholic illness—between real and false squirts—now stands seemingly outside modern jurisdiction. But if all cases of "melancholy illness" are dubious to us now for scientific reasons, I intend to argue that melancholy was already in question even at the period of its widest circulation, caught within contrary currents of sympathy, celebration, care, and criticism. In paintings, drama, prose works, and poetry that repeatedly stage the melancholic as an object of speculation and mystery, the philosophical problems triggered by doubts about the material support for melancholy are demonstrably available to early modern people. Indeed, these works cannot be understood without an epistemologically framed account of how melancholy joins together its sufferers and its witnesses into tenuous perceptual communities, and how the enabling moments of ascription and recognition that form these links can also be undone from within.

Melancholy's overdetermined status as the object of fascination, quasi-belief, and skepticism is responsible for its curious cultural afterlife long after the demise of its physiological basis. Surviving into the present moment of panic attacks and mass medication, I intend to argue that the interpretive dynamic occasioned by melancholy persists in contemporary debates within psychoanalysis, affect theory, and philosophy about the meaning of emotional experience at the contested borderline between biological and discursive explanation.

"A Melancholy of Mine Own"

Though often theorized in relation to lack and loss, melancholy might be better grasped as a problem of discursive surplus: If particular melancholy persons supposedly suffer from a lost object of love or a constitutive lack within their ego, the crowded scene of melancholic representation tends to be characterized not by the disappearance of meaning but by its manic overproduction. At once a form of madness, a sign of genius, a symptom of sickness, and a fleeting mood of sadness, melancholy made itself available to the men and women of early modern England in a promiscuously variable cluster of modes, some overlapping with one another, some contradicting one another. Along its circuitous path from fourth century B.C.E. Empedoclean metaphysics and Hippocratic medicine, through Platonic and Aristotelian philosophy, into eleventh-century Greco-Arabic medical translations and commentary, and on to its neo-Platonic Ficinian revival and subsequent importation from continental Europe, this condition had acquired centuries of intellectual prestige and a strategically flexible looseness of definition by the time it arrived onto the stages, madhouses, consulting rooms, and printed pages of sixteenth- and seventeenth-century England. This strange mixture of cumulative richness and fashionable novelty is conveyed brilliantly by Jaques in *As You Like It*:

> JAQUES: I have neither the scholar's melancholy, which is emulation, nor the musician's, which is fantastical, nor the courtier's, which is proud, nor the soldier's, which is ambitious, nor the lawyer's, which is politic, nor the lady's, which is nice, nor the lover's, which is all these; but it is a melancholy of mine own, compounded of many simples, extracted from many objects, and indeed the sundry contemplation of my travels, in which my often rumination wraps me in a most humorous sadness. (4.1.10–18)[4]

Jaques' remark that against all odds, "it is a melancholy of mine own" exemplifies the melancholy utterance as the performance of an interiority for an onlooker which oscillates between sincerity and affectation, turning identity into a thing of discursive fashion only to insist that something incommensurably personal lies behind the stereotypical surface. Breezily taxonomizing the broad number of available modes and styles of melancholy and connecting each to a particular, interested party's attempt to accrue cultural capital, impose a rhetorical effect, or consolidate some kind of power, Jaques seems to flatten melancholy into a mere placeholder of culturally mediated emotional fashion, a standard tactic in the social games played by scholars, musicians, courtiers, soldiers, lawyers, politicians, and ladies. Yet in the very teeth of this weary deconstruction of melancholy as a received idea, Jaques nonetheless insists upon presenting for our inspection a melancholic identity which is resolutely, irreducibly "his own." The simultaneous flatness and depth of Jaques' gesture define melancholy's capacity to dissolve into a seemingly endless list of socialized variants, or to solidify into an inscrutably hard kernel of self-grounding ipseity. It is this overdetermined flexibility, this capacity to inclusively expand or reductively dissolve, which I take to be constitutive of the peculiar dynamic in place between melancholy and knowledge within certain exemplary early modern English artworks. As Douglas Trevor aptly puts it, "melancholy perpetually names itself because it can never delimit its own affects."[5] That is, melancholics greet those who would know them with a curious combination of solicitude and resistance; the melancholic always seems to both require and exceed explanation, at once to need no introduction and to never be able to stop introducing himself.

But Jaques' speech is most revealing in what it does not say. Though he passingly describes his state of mind as a "most humorous sadness," we do not find here a specific reference to the substance of black bile itself. This speech is not a diagnostic report about a physiological imbalance, an excess of *melaina-kole* for which Jaques seeks a therapeutic cure. Nor do we find a reference to the melancholic as a particular temperamental type, a specific kind of person who is inherently and essentially distinct from phlegmatic, choleric, and sanguine persons. Instead, melancholy is understood as something that can be recognized and categorized across a wide range of social types (one senses that Jaques could expand upon the list of scholars, soldiers, courtiers, lawyers, ladies, and lovers indefinitely). Even at the individual level, melancholy is already plural: Jaques recognizes melancholy in himself as a hybrid construction formed out of disparate, found sources, the "objects" and "simples" out of which he has self-consciously formed this "melancholy of mine own." Jaques' speech thus dematerializes and deindividualizes melancholy. Neither a singular kind of elemental substance nor a singular kind of subject, Jaques understands melancholy as a multiplicity, an expressive array of materials and postures and cases distributed across the social surround. *In other words, Jaques apprehends melancholy as an assemblage.*

From Early Modern Assemblance to Deleuzian Assemblage

Summoning images of eighteenth-century mechanism and twentieth-century modernism, the word *assemblage* would seem to stand at a certain distance from the sixteenth- and seventeenth-century archives that are the subject of this book. The Oxford English Dictionary (*OED*) confirms that early moderns preferred the now-obscure *assemblance*, as when Falstaff contemptuously asks "Care I for the limb, the thewes, the stature, bulk and big assemblance of a man? Give me the spirit, Master Shallow."[6] The *OED* editors allege that this means simply "appearance," with the faintly pejorative sense of "show" lurking behind this tension between surface and spirit, suggesting that "assemblance" in this case derives from the French verb "sembler" and the barely English loanword "semblable," memorably derided by Hamlet in his sparring with Osric: "to make a true diction of him, his semblable is his mirror."[7] But "assemblance" can also mean the body as a corporate gathering of physical parts, and it can open out to indicate an assembly, a crowd, a gathering of distinct bodies, as when Artegall spies a crowd and wishes "to weete the cause of their assemblaunce wide" in Book Five of Spenser's *Faerie Queene.*[8] The closest surviving relative to *assemblance* is, of course, *assembly*, with its connotation of a political multitude, but in early modern texts we find assembly sometimes truncated to the noun *assemble*, where it can designate "the coming together of two persons or things; meeting, conjunction, union," as when Caxton's English printed edition of the *Golden Legend* (1483) speaks of "the unyte and assemble of the flesshe of oure lord and of oure lady."[9] Within the polysemous matrix of the early modern lexicon, material bodies can form an assemble/assembly/assemblance/assemblage which connotes an individual body, the formation of a concept, the consolidation of a crowd, and alternately sexual, religious, and poetic fusions of flesh and spirit.

Sauntering across the boundary between interiority and exteriority, by the eighteenth century the word designates both groups of physical objects and psychological aggregates: If Locke can allege in 1704 that "All that we amass together in our thoughts is . . . the assemblage of a great number of positive ideas," one hundred years later Martineau sees nothing particularly metaphoric in describing a French chestnut grove as an "assemblage of bare poles."[10] This flexible capacity to refer to both intentional structures and natural formations softens an increasingly vexed division between mind and world, and it is unsurprising that the term is widely adopted by visual artists whose work hopes to soften that boundary. In the critical art practice of the 1960s, the word becomes synonymous with the now-ubiquitous technique of collage, designating a medium unto itself minimally defined as the construction of a new form out of constituent parts, with no commitment to any particular material format. Increasingly mobile, the word stretches to fit the "combines" of Robert Rauschenberg, the repurposed thrift store *objet trouvé* of Joseph Cornell, and the agitprop theatrical tableau of Ed Kienholz. It also loops backward to include a pantheon of Dadaist precursors and cuts sideways to rope in Happenings and Event Art theorized by Allan Kaprow and others. Such variability is a function of the increasing strain under which particular mediums mutually poach, compete, and blur. Assemblages are formed out of multiplicities, represent the world as inherently plural and capacious, and take multiple forms accordingly.

Building upon the simultaneously constructive and solvent effects rehearsed in these etymological roots, the word "assemblage" has since taken on an ambitious load-bearing capacity in the collaborative texts of Gilles Deleuze and Félix Guattari, and functions as a handhold and a keyword in *A Thousand Plateaus* (1980). In a much-discussed passage in their "Postulates of Linguistics" chapter, the word "assemblage" occasions an expansive vision of bodies and signs as a dialectical manifold:

> On a first, horizontal, axis, an assemblage comprises two segments, one of content, the other of expression. On the one hand it is a *machinic assemblage* of bodies, of actions and passions, an intermingling of bodies reacting to one another; on the other hand, it is a *collective assemblage of enunciation*, of acts and statements, of incorporeal transformations attributed to bodies. Then on a vertical axis, the assemblage has both *territorial sides*, or reterritorialized sides, which stabilize it, and *cutting edges of deterritorialization*, which carry it away.[11]

At the risk of rehearsing long familiar critical terminology, permit me to distinguish the components of Deleuze's abstract schematic that I intend to make portable for the task of understanding early modern melancholy. Parsed through the antinomy between content and expression, a bodily realm of actions and passions is tethered to its complement, the semiotic realm of enunciations and statements. This relational structure constitutes the assemblage in its horizontal dimension as an already-corporate interface between two subsidiary assemblages: In a manner reminiscent of the project of Michel Foucault, we can say that the machinic assemblage of "bodies" and the collective assemblage of enunciation (or "discourse") together articulate a framework for thinking about the way that ideology acts upon material bodies and the way that material bodies, in turn, discipline themselves into forms susceptible to such recognition or pull against and distort such alignment.[12] Another way to summarize this horizontal framework is to note its resemblance to what Alan Badiou has termed the contemporary conviction of *democratic materialism*: the axiom that "there are only bodies and languages."[13]

But Deleuze and Guattari's account of the assemblage goes beyond this two-term alignment, for the entire complex structure captured by that horizontal opposition is then multiplied by a second, vertical axis of stability/instability, thus subjecting the assemblage to spatial drift and temporal change, to processes of emergence and transformation which cannot be foreseen and which produce immanent self-difference. Anything that reinforces the stability of an assemblage as it occupies space over time tends toward its territorialization. Anything that tends to dissolve, soften, distort, or change that stability constitutes its deterritorialization.

The fluidity of such a formulation is both its strength and its flaw. In response to the hapless question "If there is no single field to act as a foundation, what is the unity of *A Thousand Plateaus*?," Deleuze is direct:

> It is the idea of an assemblage (which replaces the idea of desiring machines). There are various kinds of assemblages, and various component parts. [. . .] In assemblages you find states of things, bodies, various combinations of bodies, hodgepodges; but you also find utterances, modes of expression, and whole regimes of signs.[14]

The word "hodgepodge" risks giving the game away because such elasticity courts meaninglessness. If, as recent work in Deleuze studies demonstrates, the term *assemblage* can be deployed at every level of scale from atoms to vertebrate microfossils to bodies to social collectivities, then what is added to any particular level by pointing out that here, too, lies an assemblage?

It is here that we would do well to recall that Deleuze regarded himself as an empiricist: Everything is an assemblage, but each assemblage is distinct from every other assemblage, and individual assemblages manifest their distinct expression in ways that can be known. Though they are provisional, assemblages differ from one another in ways that nonetheless generate criteria and consistency:

> Assemblages exist, but they indeed have component parts that serve as criteria and allow the various assemblages to be qualified. [. . .] An assemblage is first and foremost what keeps very heterogeneous elements together: e.g. a sound, a gesture, a position, etc., both natural and artificial elements." The problem is one of "consistency" or "coherence", and it is prior to the problem of behavior. How do things have consistency? How do they cohere? Even among very different things, an intensive continuity can be found.[15]

Each assemblage manifests a specific, local coherence and its own consistency as a function of its coding, or territorialization. Indeed, without that, one could not identify the presence of an assemblage in the first place. Because an assemblage is relational and volatile, it is not grounded in an unchanging essence that permanently fixes its identity, but just because identities are provisional and incomplete does not mean that recurrent patterns, and the criteria of assessment that go with that recurrence, do not exist. The coding and territorialization of an assemblage sustains its consistency across the ceaselessness of its own becoming—until an assemblage eventually disarticulates and collapses, or, in other words, until its subpersonal components get reassembled into other forms and relationships. The dynamic tension between territorialization and deterritorialization simultaneously grants a certain criteria of identity to a given assemblage while building the reality of change, process, and movement into how that identity might be thought.

Joining structure to fluidity, the assemblage designates both temporary socially extended relations (e.g., the assemblage of persons attending a performance of a play) and the more persistent fixity of artistic constructions (e.g., the formal fiat set in place within, say, a Kurt Schwitters collage).[16] Sometimes they show up as emergent complexities that are not tethered to any particular creative source (watches without watchmakers). Sometimes assemblages strongly index a particular creative agency, even and especially when they are formed out of constituent content taken from other sources (Jaques' creative pride in fashioning a "melancholy of mine own" out of found materials thus reiterates Ron Silliman's dictum that "Collage is a false democracy.")[17] Here we would do well to consider the French word which is translated into English as "assemblage," *agencement*, with its flavor of activity and personalized agency, and its spiral of associated synonyms: layout, organization, disposition, combination, composition, liaison. Thus, the word joins highly disparate exempla. Bodies are assemblages, but so are texts. Symptoms are assemblages, but so are audiences. Assemblages describe ongoing, fragile, and reversible correlations between bodies and languages.

Melancholy × Assemblage = ?

Having located "assemblage" within its larger family tree of signification and parsed its Deleuzian meaning, what would it mean to think of the wildly divergent and multiple phenomena of early modern melancholy as an assemblage?[18] Within the chapters that follow, the term *melancholy assemblage* fastens onto a range of disparate conceptual phenomena, distinct "cases": the positioning of the body into a temporary posture which expresses melancholy affect as a conventional pictorial form; the socially extended networks of spectatorship, diagnosis, and witnessing in which melancholy bodies and minds are recognized and interpreted by others; and the syncretic texts in which melancholy is structured into a distributed body of knowledge as an assemblage of various voices, sources, stances, and authorities.

The phrase "melancholy assemblage" must be understood as a site of theoretical crosstalk in which two signals modulate each other. Treating melancholy as an assemblage rather than a type of substance or a type of subject breaks its conceptual unity into an extended, provisional, and modular set of relations between and across material elements, and relationships between and across individual subjects, with no particular local expression enjoying any particular ontological priority over any other. To see this, let us briefly return to the quotation from *As You Like It*. Looking outside himself at the social surround, Jaques depicts melancholy as a flowing current which alters as it distributes itself through universities, barracks, courts, courtrooms, and boudoirs, generating a wide repertoire of local affective forms that may or may not resemble one another (pride, emulation, ambition, niceness, etc.). Looking within himself, Jaques encounters even his own irreducibly singular melancholy as an assemblage constructed out of multiple parts in a manner that is part pharmaceutical "receipt" (or recipe) and part artistic composition. This passage from outside to inside and back again suggests that Jaques already realizes an essential principle of assemblage: Assemblages can be stacked within one another, moving up and down scalar levels. Jaques' particular melancholy body, itself already an assemblage, is also a subcomponent of a larger melancholy assemblage distributed across social space. For a visual emblem of this principle, consider the charged cluster of two figures within an early depiction of melancholy from the Augsburg Calendar.

Seemingly indifferent to each other and yet slumped suggestively beside each other, the human figures are no more nor less melancholy than the solitary distaff and the curiously book-like table that flank and support them: elements temporarily aligned in a charged affective space in which minds, bodies, and things melancholize together. This field of multiplication between individuals and across space, minimally present

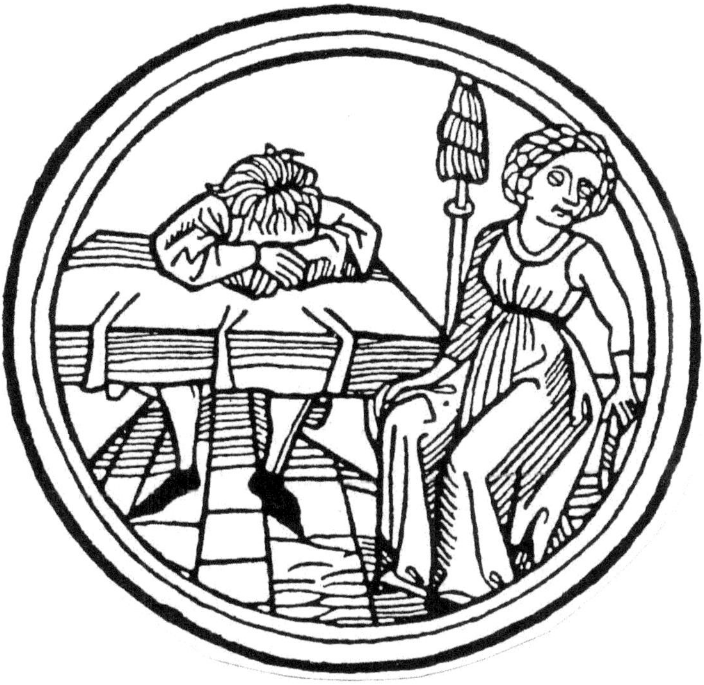

Figure 1. Anonymous, *Melancholy*, Augsburg Calendar, circa fifteenth century.

in the Augsburg Calendar, thickens and intensifies in an unsigned image by Albrecht Dürer, *Der Verzweifelnde* (*The Despairing*, 1515), in which five figures are compacted together into a claustrophobic constellation of seemingly incongruous attitudes: attentive, impenetrable, enraged, stupefied, and asleep.

Rarely discussed, this image abides in an art-historical abyss, wedged between contradictory interpretations as either a free flowing experiment or a highly self-conscious humoral allegory. As Leo Strauss notes, "There

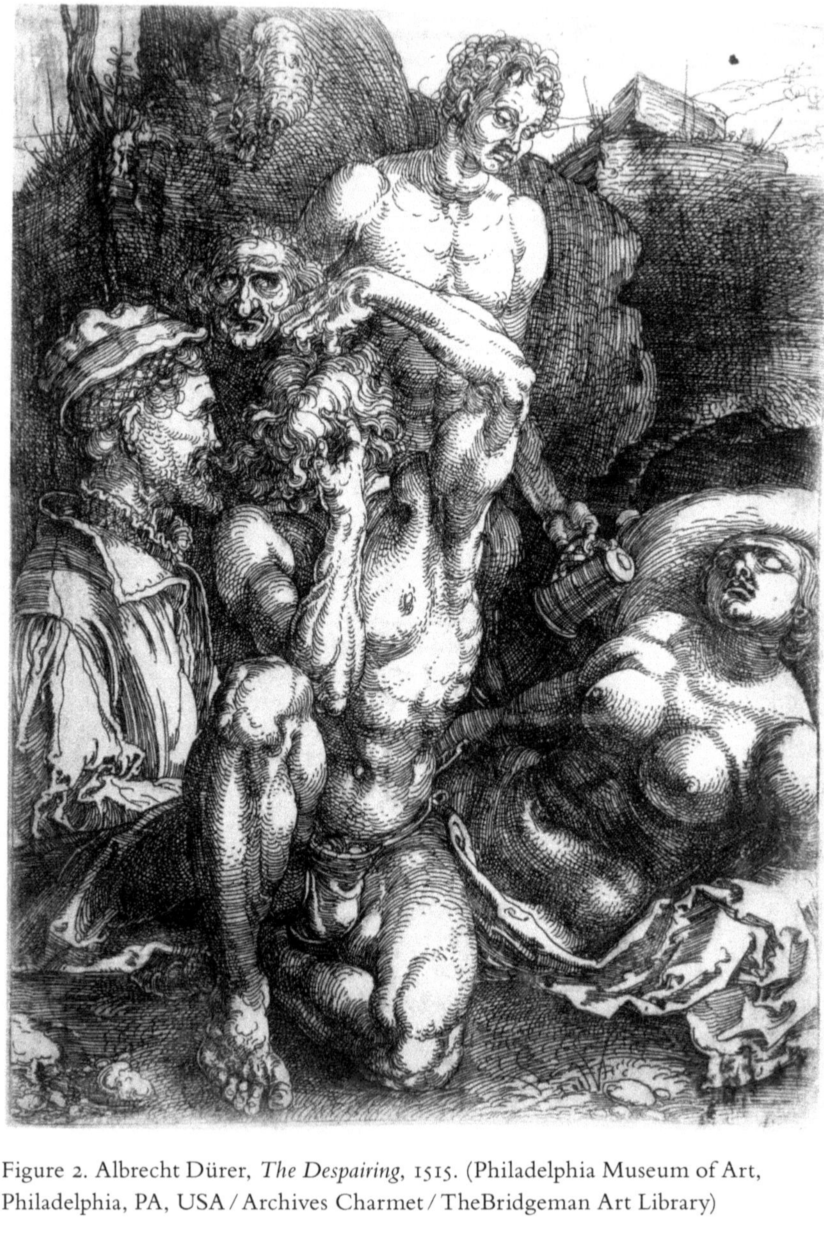

Figure 2. Albrecht Dürer, *The Despairing*, 1515. (Philadelphia Museum of Art, Philadelphia, PA, USA / Archives Charmet / TheBridgeman Art Library)

is today general agreement that this is Dürer's first experimental etching. It is unsigned, undated, and probably drawn directly on the plate without a definite program."[19] Yet, for Erwin Panofsky, a quite definite program drives the image; "The Despairing" taxonomizes the various pathological symptoms of "melancholia adusta" and asks to be read diagnostically (the sleeping woman is a phlegmatic melancholic, the central figure is a choleric melancholic, etc.).[20] At once experimentally loose and iconographically loaded, Dürer's beguilingly unstable image compacts affective flows and epistemological barriers into a sticky, self-differential engine of feeling and meaning. While I shall dwell at length upon the visual archive of melancholy in the next chapter, what interests me here is simply Dürer's rearticulation of the theoretical implications already present in the homelier example of the Augsburg calendar image. Melancholy emerges in each figure as, à la Jaques, "a melancholy of their own," yet melancholy also abides precisely in and as a plural social and material assemblage of bodies being together.

In both its intrasubjective and intersubjective dimensions, then, melancholy is an assemblage, and this plurality modulates how we understand the term *melancholy*, taking away its connotations of solitude and interior essence in favor of a model based on social extension. Put crudely, melancholy goes from being something that shows up first and foremost "in here" to being something that is always also "out there." But currents can flow in more than one direction. It must be said that "melancholy" also modulates and transforms "assemblage" just as radically as "assemblage" modulates "melancholy" from unity to plurality, for it introduces a repertoire of philosophical problems and stances which are typically absent from the political and metaphysical concerns of assemblage theory: problems of verification, truth, and knowledge. The author-function that regulates what is *Deleuzian* about assemblage theory would seem strongly predisposed to slant the term toward the monist pole of assemblages as merely expressive modalities of an essentially undivided substance. That author-function is often set in opposition to the bugbear of dualism associated with epistemology, which the recent critical flourishing of object oriented philosophy, speculative realism, and social ontology has only brought into further disfavor. Consider the sarcasm in Graham Harman's summary of correlationist metaphysics as "the tear-jerking modern rift between the thinking human subject and the unknowable outside world."[21] Harman's polemics repeatedly harp upon gloom, desolation, or despair as the affective character of the epistemological stance saddled with the unresolvable split at the core of dualist metaphysics, the "magical gap between thinking, practical, moody humans on the one hand and stupefied inanimate clods of matter on the other."[22] Though Harman is discussing a rift between Latour and Kant in relation to their metaphysics of substance, his rejection grounds itself in an intuition that the epistemology of gaps and the affect of despair are mutually tethered.

Deleuze arguably anticipated this current hostility to epistemology. In much of his own work without Guattari, Gilles Deleuze articulates a "univocity of Being" whose monist flavor would seem predicated upon a rejection of the epistemological stance in favor of a material or substantial form of empiricist realism, a stance that might now be described as a "flat ontology" and aligned with a range of new developments in metaphysics which attempt to bypass phenomenology, finitude, and epistemological gaps in favor of immanence, emergence, and object-oriented accounts of being.[23] Deleuze's Spinozist repudiation of "sad passions," his reach in his final writings toward "complete power, complete bliss" would seem to stand at a salutary, polar opposition from the melancholic nadir of crippling despair. He is the thinker least prone to seduction by melancholy as a tone of voice, and in that sense provides a usefully estranged critical vantage point upon its contaminating power. Deleuze's elegiac final celebration of pure immanence marks the limit point of his attempt to think beyond the categories of subject and object upon which epistemological questions depend: "Absolute immanence is in itself, it is not in something, *to* something; it does not depend on an object or belong to a subject."[24] But Deleuze's own last word need not be ours.

Bracketing the question of whether Deleuze and Guattari regarded assemblages as a way of talking about the world that would sidestep epistemology, in christening "melancholy assemblages" I am treating the epistemological stances, subjects, and contents of early modern melancholy as components that are connected in complex relations with objects, signs, and bodies. As such, they are not only susceptible to inclusion within assemblages but, I shall argue, can be best understood within that rubric. If such a hybridizing gesture risks granting a concession to precisely the paradigm that Gilles Deleuze in particular sought to escape in the first place, I am afraid that I will need to considerably expand my account of the early modern historical circulation of melancholy in order to explain why this is a risk worth taking and, indeed, why melancholy cannot be understood outside of a certain liminal relationship to the epistemological stance.

The Epidemiology of Melancholy Assemblages

The story of the cultural history of melancholy as a sequence of vectors and arcs of discursive transmission has been told often, and well.[25] If the "just so" story of "How Melancholy Got Its Cachet" constitutes a much-rehearsed narrative within Renaissance intellectual history, my goal in retracing this trajectory is not simply to produce what William James termed "a monotonous rain of instances" but to reconfigure melancholy in the process.[26] A polychronic assemblage territorialized around a core set of author-functions (Empedocles, Hippocrates, Aristotle/Theophrastus, Galen, Ficino), which cumulatively develop an expressive pattern, a symptomatology, and a stable set of interpretive practices and habits, melancholy ultimately ran aground upon a central contradiction whose consequences remain undertheorized: the ongoing tension between the Aristotelian/Theophrastan account of genial melancholy and the Galenic account of pathological imbalance. This tension disorganizes the concept but also paradoxically introduces a secondary kind of consistent incoherence, or generative indeterminacy, into the expression of melancholy itself. This conflict wedges a kind of persistent kink in the conceptual field of melancholy, a distortion that drives the startling overproductivity of melancholy meaning and melancholy representation relative to the other humors and temperaments within their shared humoral materialist scheme.

"Patient Zero" is Empedocles, whose visionary cosmological poem announces a quadratic elemental metaphysics with a profound influence upon the course of Greek medicine. We only possess fragmented quotations of this lost text, but the extant fragments articulate a model of the universe as inherently multiple: Earth, air, fire, and water constitute what Empedocles terms *roots*—elemental components of a material ontology.[27] Rather than simply declaring that "there are only four kinds of stuff" within a material universe, Empedocles' text insists that the complex fact of mixture and intermingling goes all the way down: Any particular object is itself already compounded of different ratios between these four materials, and those relationships are dynamically unfolding and changing over time as distinct forms emerge, die, and transform.[28] Over and above the notional purity of these four elemental "roots" themselves, there are only manifold ways of being more or less airy, fiery, earthy, watery. These states are inherently kinetic and subject to change because, for Empedocles, all material mixtures are constantly acted upon and transformed by the rival animating forces of Love (a territorializing force that binds material mixtures together) and Strife (a deterritorializing force that dissolves bonds and collapses organizing mixtures).[29] Accordingly, humoralism has its basis not in a moment of "normal science" but in a lost object on the horizon line before the disciplinary segregation of poetry and philosophy: Empedocles' missing master text stages a visionary psychomachia of Love and Strife as rival, dynamic forces in which the world itself is a material assemblage constantly under (re)construction.[30]

Hippocrates produced a medical science in the image of this Empedoclean cosmology. The two coordinating polarities of Hippocratic medicine, heat versus cold and dryness versus moisture, are overlaid onto the Empedoclean schema of earth, air, fire, and water, and the result is the medical doctrine of the "four humors" as somatic expressions of this fundamental design principle. Blood, phlegm, yellow bile, and black bile are coordinated along axes of temperature and moisture into a dynamic tetrad of humoral psycho-physiology:

Humor	**Quality**	**Season**	**Temperament**
Phlegm:	(Cold and Wet)	Winter	Phlegmatic
Blood:	(Hot and Wet)	Spring	Sanguine
Yellow Bile:	(Hot and Dry)	Summer	Choleric
Black Bile:	(Cold and Dry)	Autumn	Melancholic

In the transition from Empedocles' lost source to Hippocratic prognostic texts such as *On the Nature of Man*, anatomy replicates cosmology: The body is the site in which all four liquid *humon* (humors) flow, surge, interact, and transform themselves.[31] Balance (*eukrasia*) produces health, while imbalance or excess (*dyskrasia*) produces illness.

Literalizing melancholy's status as a "disease of the mind," let me briefly fast-forward through the viral narrative of discursive epidemics, hermeneutic panics, historicist cures, artistic mutations, and new theoretical outbreaks which follows from this Empedocles-to-Hippocrates moment of "transmission." Infected by Empedocles with an ontology of mixture and assemblage, Hippocrates transmits the doctrine of melancholy disease to Aristotle, who transmits the disease to his immediate disciple Theophrastus and to his later inheritor Galen of Pergamon, who each mutate the meaning of melancholy by producing highly influential, mutually incompatible definitions. Thanks to an error of attribution in which the notes of the pupil are mistaken for the doxa of the master, Theophrastus' text "Problemata XXX.1" enters the philosophical canon as one of the authentic works of Aristotle. It opens with a notoriously leading question, an unsubstantiated premise that casts a long shadow across melancholy's subsequent intellectual history: "Why is it that all those who have become eminent in philosophy or politics or poetry or the arts are clearly of an atrabilious temperament, and some of them to such an extent as to be affected by diseases caused by black bile, as is said to have happened to Heracles among the heroes?"[32] Defined as a genial gift from the gods, Theophrastus' "Aristotelian" form of melancholy designates a temperamental trait associated with exceptional excellence in a dazzlingly wide range of human activities, gathering a pantheon of exceptional heroes from philosophy and politics and poetry and the arts into the most enviably select "melancholy assemblage" of them all. It must be said that even for this hybrid author Theophrastus-Aristotle, melancholy must be managed carefully so as to avoid violent excess, in keeping with broader Aristotelian commitments to the ethical regulation of the passions and the maintenance of the mean. But out of all of the humoral temperaments, the predilection to melancholy is something to be cultivated in a crucially ambiguous sense: something to be handled so that it does not overstep its bounds, but also something to be nourished, even celebrated. Crucially, a lack of melancholy disqualifies a person from attaining the heights of human excellence.

Starkly opposed to such a reading, for Galen melancholy is simply one of the four possible humoral imbalances implicit in Hippocratic physiological pathology, and hence not a potential index of personal excellence but simply a physical illness with mental side effects that physicians ought to try to cure. This restricted sense of melancholy's significance had wide consequences, for Galen of Pergamon's writings on melancholy formed a corpus of eclectic commentary upon Hippocratic ideas that defined the standard theory and practice of how to diagnose and treat melancholy for most early modern medical practitioners.[33] Drawing a strong distinction between melancholy that originated in the body (deriving from imbalances or excessive heat in the blood, stomach, or liver) and melancholy that originated in the brain, Galen proposed distinct therapeutic approaches for these separate forms. What united his analysis of the condition across such distinctions was his oft-quoted definition of the fundamental symptom of melancholy as fear and sadness: "Although each melancholic patient acts quite differently than the others, all of them exhibit fear or despondency."[34] The melancholic as Galen depicts him is afflicted with colorful and exotic delusions and emotional and mental symptoms (paranoid fantasies, baseless anxieties, delusions of grandeur) as well as humiliating physical ones (sallowness, flatulence), but this person is essentially a sick human being to be treated with bloodletting, herbal remedies, cooling baths, or purgatives. Galen's melancholic is a resolutely mundane creature, a miserably unhappy sufferer of a physiologically understood disease, but no more than that. Helped along by successive waves of translation into Latin, the standardization of canonical authority in the early universities, and the explosion of print technology, Galen succeeds in transmitting his physiological definition of melancholy into the bloodstream of the entire early modern medical profession. Galen becomes the standard textual authority for the hegemonic medical doctrine of humoral materialism.

If the Galenic definition of melancholy reached the most people, the Theophrastan-Aristotelian definition reached the "best" people. This philosophic strain infects Marsilio Ficino and Petrarch, and their commitment to the genial understanding of the disease induces a chain of epidemic outbreaks in the humanist sodality of the Florentine Academy and, eventually, the urban centers of Renaissance Italy. Melancholy becomes the consequence of, and evidence for, commitment to a scholarly ego ideal: a sign that one was bookish, profound, withdrawn, learned, elite, aloof, and ill. The titles of typical chapters from Ficino's *De Vita Libri Tres* [Three Books On Life] (1489) ("Learned People are Subject to Phlegm and Black Bile," "How Many Things Cause Learned People Either to Be Melancholy or to Become So," "Why Melancholics Are Intelligent") not only reinforce the Theophrastan doctrine of melancholy but make it contagious through acts of reading.[35] Expanding the range of membership in this growing social cluster, Petrarch thickens the demographic of potential subscribers to the Theophrastan doctrine by rendering melancholy the signature affect of distracted, sorrowful romantic passion. Alternately scholarly and amorous in its inflection, this civilizing Italian form of fashionable melancholy infects English travelers to the continent, who import the Aristotelian-Ficinian hybrid form of the disease into early modern London. Upon arrival, it quickly spreads throughout the urban culture of the capital, entering into circulation in sermons, ballads, poems, verse satires, and on the public stage as both a tragically serious condition and a comedic target of satirical exposure.[36] The outcome is overproduction: a civic surplus of melancholy self-representation.

Jonson illustrates the resulting ubiquity of melancholy in a biting scene from *Every Man in His Humour*, when the "country gull" Master Stephen meets the "town gull" Master Matthew and makes a point of ostentatiously announcing himself as a sufferer of melancholy illness. What ensues is a deeply satirical cycle of competition disguised as solidarity:

> *MATTHEW:* But are you indeed, sir, so given to it?
> *STEPHEN:* Ay, truly, sir, I am mightily given to melancholy.
> *MATTHEW:* Oh, it's your only fine humour, sir; your true melancholy breeds your perfect fine wit, sir. I am melancholy myself divers times, sir, and then do I no more but take pen and paper presently, and overflow you half a score of a dozen of sonnets at a sitting.
> *EDWARD KNO'WELL:* (aside) Sure he utters them, then, by the gross.
> *STEPHEN:* Truly sir, and I love such things out of measure.
> *EDWARD KNO'WELL:* (aside) I'faith, better than in measure, I'll undertake.
> *MATTHEW:* Why, I pray you, sir, make use of my study; it's at your service.
> *STEPHEN:* I thank you, sir; I shall be bold, I warrant you; have you a stool there, to be melanch'ly upon?[37] (3.1.79–91)

In a humoral pandemic, melancholy produces literary representations which inspire performative imitations of that emotion that themselves inspire further literary productivity. Squirting "false" copies of itself onto the page and onto the world, melancholy is not a substantial medical condition but a viral machine caught in a loop of auto-replication, a discursive phenomenon that induces the enactment and recirculation of a set of clichéd statements, postures, symptomatic poses, and attitudes which are instantly legible but never verifiable. Hitting below the belt, in having Stephen ask for a "stool" upon which to be melancholy, Jonson renders the resulting poetic outpouring of melancholy feeling into a not-so-subtle equivalent for the child's fecal "gift" to the world ("stool" is a common contraction for "close stool," or commode).

What stance is being taken here toward the material support for humoral illness? In lampooning the earnest declarations that these gulls feel "true" melancholy, is Jonson taking the further step of alleging that there is, in the words of Gertrude Stein, no *there* there for them to feel in the first place? Logically, cases of false melancholy illness do not prove that real cases of melancholy illness do not exist, and urgent satirical policing of "false positives" might be made on behalf of a respectful sense of the severity of genuine melancholy illness. In *Humoring the Body*, Gail Kern Paster argues that "Jonson's satiric attack on the humors in plays such as *Every Man in His Humour* [. . .] is not, as is often supposed, an assault against a waning humoral belief system but a critique of socially performative uses of the humors in order to flaunt eccentricity or license unwarranted aggressivity."[38] However such a reading may chime with the self-understanding of Jonson's ethical project as announced in his prologues and with Kno'well's invocation of the just "measure" which melancholics exceed, it only begs the question of whether there could be moments of humoral self-description which would not be "socially performative" in this sense. More importantly, such an account hardly explains why the problem of verification, of determining whether one has a "true" or "false" condition, should attend melancholy specifically, when one cannot discover similar debates about whether characters are really choleric, really sanguine, or really phlegmatic. Why was melancholy specifically subject to epistemological examination in a way that other humors were not?

Jonson's scene suggests that melancholy came to be "an epistemological monster" as a result of the centrifugal effect that its own popular circulation had upon the basic conceptual conflict between its two rival definitions: Master Stephen and Master Matthew aspire to the Aristotelian inflation of the illness, while Kno'well (and through him Jonson) comically assert its Galenic reduction. In *The Elizabethan Malady*, Lawrence Babb, one of very best practitioners of what, without condescension, one would want to call the Old Historicism, tidily identifies this very tension as melancholy's defining paradox:

> The Renaissance, then, held simultaneously two conceptions of melancholia. According to Galenic tradition, melancholy is a most ignominious and miserable condition of mind; according to the Aristotelian tradition, it is a most enviable and admirable condition of mind. Scholars in general denied the truth of neither; for behind the one was the authority of Galen and behind the other the authority of Aristotle. These two conceptions are hopelessly entangled in Renaissance thought and literature. Sometimes they seem at least partially reconciled through the nice distinctions of the psychologists; sometimes they seem very much at war with each other.[39]

How could both Galen and Aristotle be correct? How could melancholy be both a numbingly ordinary medical illness and the sacred sign of divine favor and personal excellence? Surely this blatant contradiction must have produced a range of responses from early modern subjects keen to absorb and apply classical learning to their own experience. Yet at the end of *The Elizabethan Malady*, Babb dismisses the distinction with a whimper:

> Englishmen were not troubled by the opposition between the two concepts of melancholy. They accepted both. Yet on the whole the more dignified connotations of melancholy determined England's attitude toward it. In general the attitude was definitely one of respect.[40]

Recent critics have not shared Babb's genteel view of the matter, but their responses have overcompensated for that fault while sharing the same basic either/or architecture. Where Babb rosily asserts the dominance of the Aristotelian definition of melancholy, more contemporary accounts of the condition—such as Gail Kern Paster's *Humoring the Body* (2004) and Douglas Trevor's *The Poetics of Melancholy in Early Modern England* (2004)—have swung in the opposite direction, steadily re-Galenizing the disease. There is a sound and simple reason to do so: Galen's firm clinical preeminence over Aristotle in the everyday diagnostic situations in which medical practitioners would have consulted with patients. Conceding the victory to Galen on structural and pedagogical grounds, both Paster and Trevor defuse the broader cultural opposition between rival definitions of melancholy and largely sidestep the resulting debate, in the process reifying Galen into a univocal position of "intellectual dominance."[41] Pace Paster and Trevor, I see less dominance and more conflict. Though the critical conversation about melancholy has gained sophistication as it has taken on board new vocabularies and new concerns, Babb's essential insight—that melancholy constituted an intellectual battleground between genial and pathological understandings—has gone missing. Babb himself is partially to blame for the ease of this disappearance. Babb detected the opposition within the melancholic archive but preemptively deflated its significance in arguing that the essential and fundamental difference between Aristotelian melancholy and Galenic melancholy had no particular consequences.

This book will argue that Babb was right about the conflict but wrong about its result. English people *were* troubled by this very opposition, and the often startling literary and artistic and social assemblages that they fashioned in response constituted a means of working out the consequences of this foundational disparity. Countering the quietism of Babb and the unilaterally Galenic readings of Paster and Trevor, I suggest that the push-and-pull between melancholy's rival definitions was already culturally available within the early modern period itself, and that the epistemological problem of how to understand melancholy across the divide between Aristotle and Galen was precisely the source of the concept's curiously elastic power both then and now. Far from subscribing uncritically to ancient Greek medical dogmas which determined their experience from behind a veil of classical prestige, early modern people critically activated the conflict between Galenic and Aristotelian understandings of melancholy and saw melancholy as volatile, up-for-grabs—the stuff from which art and argument and assemblages are made. Far from living in a lost golden age of fusion in which substance and affect were harmoniously connected and inseparable, early moderns worried about the distinctions and connections incessantly.

If recent critics have underestimated the lasting importance of this conflict, I do not want to overcompensate by simply reasserting the Galen-contra-Aristotle conflict as an overly rigid dichotomy, a single either/or choice. It is hard to see how such a basic contradiction could become culturally generative and productive if it was a simple matter of choosing between two clearly incompatible sides, particularly when they are not at all evenly matched in terms of scope and theoretical ambition. Ultimately, the interpretation of melancholy must go beyond the reductive task of determining simply which ambient cultural background is to be master: If we know, thanks to Babb, at least part of the reason why melancholy became an epistemological problem for early modern subjects, that still does not explain the cultural consequences of that conflict.

The Epistemology of Melancholy Assemblages

Jonson's comic *reductio ad absurdum* of melancholy's cultural ubiquity suggests that by the dawning of the seventeenth century, the primary symptom of the disease of melancholy was not "fear and sadness" but the overproduction of its own interpretation. Accordingly, if we are to understand the "problem space" of melancholy circulation within early modern culture, we must redefine melancholy as neither a substance nor an affect but at a more primitive, basic level as an assemblage predicated upon knowledge, a particular stance of recognition.[42] Though melancholy requires as grounding conditions historically inflected correlations of bodies and mental states, these are necessary but individually and collectively insufficient components when describing that condition. Though it requires those supports, those supports are not in and of themselves "melancholy"—for melancholy itself only obtains as a kind of knowledge about those supports, an ostensive stance of recognition that encounters them and binds them into a melancholy assemblage with a diagnostic decision or an ascriptive declaration. If we are to cover the range of cases which produce this sort of knowledge, it would have to be whittled down to a knowledge of a minimal, even circular kind: "you are melancholy when someone recognizes and thus knows that you are melancholy." This seems very thin indeed.[43]

Of course, it is not really a thin definition at all, for the term *recognition* already smuggles in 1) an entire theatrical dynamic of signifying emotional behaviors to be recognized externally, 2) an entire passionate repertoire of affective states to be recognized internally, 3) a notional self or consciousness to do the recognizing, and 4) in the mere "re" of its prefix, an entire temporality of origin and repetition, of presentation and re-presentation extended over time. Widened to include all of these preconditions, melancholy can be both anchored in an epistemological redefinition and capacious enough to accommodate its range of cases.

You are melancholy when someone (yourself potentially included) knows that you are melancholy. To take the normative case of "sincere" melancholy first: If I feel something and act a certain way as a result of those feelings and those actions are recognized by another as the actions associated with the inner feelings that are said to be constitutive of melancholy, then your recognition establishes my melancholy without any need for any such recognition on my part. I can be melancholy without myself recognizing these feelings and actions—effectively, without knowing that I am melancholy. Antonio's showy declaration in *The Merchant of Venice* that "I know not why I am so sad" (1.1.1), an announcement that will detain me for some time, insists upon this possibility. But I cannot be melancholy if *nobody* recognizes this and knows it because melancholy requires a kind of conscious ascription on the part of at least one observer, spectator or witness. The more witnesses, the greater the number of bonds and exterior relations joined together to consolidate the assemblage, the "stronger" is the ascriptive force of a particular melancholy assemblage.

Does this mean that melancholy is always a "public" matter? Even if the capacity to recognize and ascribe is itself public, a social and learned behavior, a given experience of melancholy ascription need not take place for others. If my own feelings or actions show up for me as melancholy, then I can recognize and know that I am melancholy without necessarily routing that meaning back into discourse for others. Melancholy does not have to be cosigned by the melancholic in question in order to happen, but it needs at least one signature from someone. That is, in both the case of someone detecting in me something that I have not detected in myself and in the case of an entirely solitary self-diagnosis, melancholy has to be knowledge in order to "happen" at all. It is thus never simply a private experience of feeling, a private or inward affective experience—it requires a kind of secondary "book-keeping" in consciousness. It is not just a feeling but a claim about the feeling, a socially extended assemblage composed out of consciousness, expression, a regime of knowledge, and at least one body.

This means that melancholy is inherently interpretive, the result of a decision, a matter of judgment. But is it therefore the *product* of a judgment? When we judge something to be melancholy, we do not think of our judgment as the source of that melancholy—we take ourselves to be noticing something about the world, not just willfully imposing by fiat. This can trigger impacted, seemingly circular debates about social constructionism, relativism, and the basis of ascription.[44] There are plenty of examples of this sort of ascriptively generated objects of knowledge which language seems to call into being, objects which are elements in a social ontology. Classical logical examples include *sorites* paradoxes concerning piles with nonspecific numbers of components: How many grains of wheat are required before one can say that there is a heap? True to form, in *On Medical Experience*, Galen relates the *sorites* paradox, and then dismisses it as a "pretty sophism," expressing a practical confidence about our ability to make perfectly successful references to heaps without collapsing into a logical abyss in the process.[45]

A less exotic form of referential hurdle must be crossed in the diagnostic experience of identifying a disease from a cloud of tokens and signals. "Heap" is vague word, but not therefore meaningless; the word refers to a collection or gathering of subcomponents whose clustering becomes significantly emergent via an act of description. The same dynamic prevails in medical discourse when a *syndrome* emerges as a pattern of organization from a broad range of symptoms. The presence of any particular symptom is not enough to constitute the disease, but if enough of the group is present, then an individual case comes to warrant the ascription of that label—but it only does so with a doctor's jurisdiction. The transition from seropositivity to "full blown AIDS" offers a case in point; the transition is produced as knowledge by a doctor who monitors the kinds of expressive symptoms and opportunistic infections present and decides when the boundary into the syndrome has been crossed.

Similarly, a range of behaviors and signs and subjective reports taken together constellate melancholy as a medical diagnosis. Simply folding one's arms or avoiding company or sighing all by themselves are not sufficient—there must be a decisive declaration to one's self or from another person that these separate signs and behaviors share a common explanation in an excess of black bile within the body that is producing the atrabilious disease. From an accumulation of grains, the heap emerges; out of a repertoire of actions and statements and postures and signs, melancholy emerges. Crucially however, those signs and behaviors can be faked. They can be caused and triggered by an inner feeling, but they need not be. The scene of melancholy recognition or ascription is therefore haunted by the possibility of falsehood and deception, a possibility that loosely tethers the phenomenon of melancholy to epistemology, neither forcing any particular example to submit to an evidentiary tribunal, but never setting any particular example decisively free from the possibility of such questioning either.

It is now time to bring the structural framework of the assemblage and the hermeneutic stance of epistemology into theoretical alignment. Taken together, the scene of Jaques' melancholic self-ascription and the scene of Dürer's affectively intense cluster of figures and the scene of Master Stephen and Master Matthew's status-seeking one-upsmanship are simultaneously located within the philosophical domain of epistemology and the formal domain of the assemblage. Insofar as such scenes and images hinge upon establishing criteria for knowledge (especially when those criteria remain frustrated or incomplete), they are inherently epistemological. Insofar as they are complex constructions of bodies and signs that are subject to historical tension between coding and decoding over time, they are assemblages. Hence, we can coin a hybrid formation that will drive the distinct readings and cases within this book: *melancholy as an epistemological-affective assemblage.*

The primary benefit to thinking of melancholy as an epistemological *and* affective assemblage is that we can preserve a respect for its remarkable consistency across its own plural set of 'examples' without needing to posit an essence that governs all cases. Doing so permits us to respect the protean nonself-sameness wedged into place by the conflict between Galenic and Aristotelian melancholy, but avoids taking a precipitous stance about melancholy's existence or nonexistence. It is its richly various diversity, even its deformity with respect to itself, that gives melancholy, paradoxically, the only expressive consistency we can impute toward it if we are to do justice to Jaques' exemplary list, the chasms of doubt in Jonson's texts, and Dürer's eloquent, tangled forms. Assemblage theory permits a kind of "realism" about the fact of multiplicity as multiplicity, which suits melancholy twice over: not only because of melancholy's nomadic trajectory across cultures and historical periods, but also because of its inherent modularity. It need not obtain at any single level of expression but can be scaled "up" or "down" as it inflects bodies, subjects, texts, cities, cultures, and nations. As Manuel De Landa puts it:

> The identity of any assemblage at any level of scale is always the result of a process (territorialization and, in some cases, coding) and it is always precarious, since other processes (deterritorialization and decoding) can destabilize it. For this reason, the ontological status of assemblages, large or small, is always that of unique, singular individuals. In other words, unlike taxonomic essentialism in which genus, species and individual are separate ontological categories, the ontology of assemblages is flat since it contains nothing but differently scaled individual singularities (or haeccities). As far as social ontology is concerned, this implies that persons are not the only individual entities involved in social processes, but also individual communities, individual organizations, individual cities and individual nation-states.[46]

Taking up this modular capacity of assemblages to "stack" within each other, let us consider the utility of assemblage theory up and down this horizontal chain invoked by De Landa. We can speak of a particular melancholy sufferer or patient as an assemblage, an incorporation of specific bodily experiences, personal memories, education, economic means, dietary regimens, self discipline, religious commitments, gender variance, and symptomatic complaints which constitutes itself in relation to a host of exterior forces while remaining irreducibly singular. But we could also speak of a particular stage melancholic (say, Hamlet) as a melancholy assemblage constructed out of a self-differential set of "subpersonal components."[47] Hamlet names not a person but a paratactic network that relays and multiplies Saxo Grammaticus and Belleforest and Seneca and Kyd and Bright and Ficino and so on; the resulting synthesis is charged with the task of interrogating its own discursivity in the living question of whether the "antic disposition" is a charade, a melancholy disorder, or a complex mixture of both "false" and "real" mental illness. Adjusting the scale of operation again, we could also think about a particular author-function as an assemblage: "Aristotelian melancholy" names not an individual doctrine but a complex assemblage of Aristotle and Theophrastus and various translators and commentators and misinterpreters and pedagogical institutions and historical accidents of transmission.[48] We could also think of London itself as a melancholy assemblage in which architecture and weather patterns and economic cycles and mortality rates and plagues and print-capitalism function together to make certain kinds of phenomenologically embodied, partial experiences of the city intelligible, portable, repeatable, and subject to change as "melancholy." Containing but exceeding the scale of that assemblage, we could then pan out to see early modern England itself as a melancholy assemblage in which urban habits and fads predicated upon such performance and recognition and satiric unveiling are disseminated and transformed and rerouted as they circulate, calcify, or succumb to regional variation across terrain, making cross-pollination between "town gulls" and "country gulls" like Master Matthew and Master Stephen possible. In each case, the scale of assembly and the relative complexity expands or contracts to include the total set of components and relations that persist across time and territory as a material and social network of forces in which melancholy affects, images, substances, and postures are formed, recognized, stored, recalled, imitated, deployed, and altered. The total set of all assemblages and their subcomponents would define the social ontology of the melancholy assemblage within early modern England as a variable, extensive, relational multiplicity of singular forms: not local examples of a unifying essence but a complex array of individual expressions.

The Contents of This Book

In each of the chapters that follow, the meaning of the phrase "melancholy assemblage" modulates and transforms itself: The result is not a linear narrative timeline of representational progress but a star-shape, a wheel with seven spokes radiating from a conceptual center. I move in a loosely chronological order from the somatic assemblage of melancholy as a bodily posture in the paintings of Nicholas Hilliard and Isaac Oliver, to the perceptual assemblage of melancholy as an extended social relation of interpretation and diagnosis in three plays by William Shakespeare (*Love's Labour's Lost*,

The Merchant of Venice, and *Hamlet*), to the scholarly assemblage of melancholy textuality in Robert Burton's *Anatomy of Melancholy*, concluding with an account of the communal assemblage formed around the melancholy aporia of Samson's martyrdom within John Milton's *Samson Agonistes*. In each case, the materials from which these assemblages are composed, and the effect of how they each locally organize and express melancholy, are correspondingly distinct; it is my hope that the result is a historically extended, cumulative multiplicity that enacts melancholy's estranging, disorganizing power.

Across this multiplicity, I hope to fashion not only "a melancholy of mine own," but also, in a kindred act of self-reflexive critical identification, "a criticism of mine own" out of diverse elements, a polychronic assemblage in which Empedocles, Galen, and Aristotle are placed in dialogue with Freud, Benjamin, Deleuze, and their rancorous theoretical aftermath. Equally attuned to unraveling overdetermined scenarios and elucidating signifying voids across the spectrum between the body and the sign, the discipline of psychoanalysis has much to offer to the task of interpreting melancholy assemblages. To be sure, the mixture of Deleuzian and Freudian registers risks distortion and misunderstanding from both sides of a lingering critical, even ideological, divide. The tensions should be sufficiently familiar to any reader of *The Anti-Oedipus* (1972), but by now that observation goes double for the subsequent generations of defensive psychoanalytic responses to Deleuze and Guattari's now-canonical critique of the socially regulatory, conservative effects of the dissemination of Freudian theory and, in particular, its American descendent, ego psychology.[49] Bracketing the divergence of Deleuzian schizo-analysis from Freudian doxa, I wager that the psychoanalytic understanding of overdetermination constitutes a means to go beyond the mere enumeration of multiplicity, to press a given surplus of meaning to determine what drives, desires, or satisfactions are at work within particular scenes of overdetermination—and for whom. But this task opens outward into historical frameworks that exceed the individual horizon of particular subjects, and it is here that Walter Benjamin's account of baroque allegory supplements the Freudian theory of the symptom with a supple combinatorial poetics strongly indebted to the genial melancholic tradition. From the saturnine perspective of the melancholic allegorist, Benjamin understands historical knowledge as itself already a dialectical assemblage of expression and convention in which "Any person, any object, any relationship can mean absolutely anything else."[50] If there is a push-and-pull between Aristotelian and Galenic understandings of melancholy within the historical archive that informs this book, there is an equally strong push-and-pull between the Deleuzian framing and the Freudian and Benjaminian readings contained within that framing. It is my gambit that melancholy's complex articulation requires such a hybrid set of critical materials.

As a simultaneously muffled and excessive affect whose primary symptom is the insistent demand for its own interpretation, melancholy comes to function as a kind of master trope for the object of criticism itself. Tracing how this came to be, these chapters articulate a shared theoretical structure through distinct historical and artistic expressions. The epistemological and verificationist crises occasioned by each particular representation within the melancholy archive stabilize the assemblage around a common effect: the hermeneutic mystery of melancholy as, in Burton's portable definition, "a kind of dotage without a feaver, having for his ordinary companions, fear and sadnesse, *without any apparent occasion*."[51] Yet they also, through the incompleteness of that very mystery, constitute the "cutting edge of deterritorialization" within the melancholy archive by eroding, blurring, and softening melancholy from within. Insofar as each expression of melancholy arrives in a "questionable shape," we can say of the melancholy assemblage that it has no center and no outliers: every point is an edge. Other than the literally essentialist material positing of "black bile" itself as a basic physiological element within the body, there can be no melancholy essence because there is not a singular kind of melancholy content. Instead, melancholy manifests itself as a characteristic affective rhythm of presentation-and-withdrawal, a blur of contradictory yet expressive symptoms (mania, depression, psychosis, dotage, flatulence, sallowness, insomnia, sexual jealousy, "lycanthropy," hallucination, lethargy . . .) whose material cause remains occulted within the body. For the bystanders who sense this rhythm of interpretive tension and release, the melancholic shows up as precisely the bearer of an affect that cannot be completely known, cannot be securely verified, and yet can be recognized and transmitted. In these moments of encounter, the territorializing stability and deterritorializing instability of the assemblage are short-circuited and become modes of each other: Melancholy thus codes through its own decoding. Rather than merely presenting us with a mystifying, sophistical aporia, it is the task of this book to treat this seeming paradox within the melancholy assemblage as the starting point of analysis.

1. *From Dürer's Angel to Harlow's Monkey*

Hanging the Head

Melancholy arises through the composition and recomposition of bodies: substances flowing, heating, and cooling within the somatic interior, limbs and extremities falling into attitudes or taking up postures, skin surfaces growing taut or slack, tiny expressive muscle systems arranging themselves into legible states of display. Over time and across culture, the smooth spectrum of bodily affect is territorialized into a striated repertoire of characteristic zones. These zones are given names and classified into emotions, a historical and material process of inscription which belies the volatility and fluctuation of its embodied support.[1] Across a chasm, presentation and representation presume and reinforce each other. We cannot create a picture of a body experiencing an emotional state without knowing what an emotion looks like in the first place. Between psyche and soma there falls the shadow of culture, which crafts conventions, taxonomies, and ready-to-hand stereotypes. The mutually reinforcing tasks of emotional recognition and emotional representation depend upon the mastery of these conventions, and presume an implicit causal circuit of expressive resemblance between interior states and exterior postures. Inner and outer are expected to move as one.

In a foundational theoretical articulation of the principles that were to guide Renaissance painting of the emotions, Leon Battista Alberti's *Della pittura* (1435) states this principle of inner/outer correspondence explicitly:

> The movements of the soul are made known by the movements of the body. Care and thought weigh so heavily that a sad person stands with his

> forces and his feelings as if dulled, holding himself feebly and tiredly on his pallid and poorly sustained members. In the melancholy the forehead is wrinkled, the head drooping, all members fall as if tired and neglected.[2]

Melancholy's legibility is simply one subsection of a broader taxonomy of passionate signs, but melancholy complicates the implied relationship between the soul and the body. For Alberti, melancholy shows up as an apparent slump in the body's capacity for energy, force, and motion. It is a slump that redirects our attention to an implied interior space of perturbed, sad passions: the soul activated by the very force that disables the body, making it at once an-expressive and extra-expressive. Neglect of the body connotes a soul turned in retreat upon itself, and the drooping head is the persistent index of this passage inward.

What is it, exactly, that we witness when we see someone as a melancholic? How did melancholy individuals show up for their early modern spectators and how do they show up for us today? To the extent that I am engaged in a redescription of the experience of witnessing melancholy, I am engaged in a phenomenological description of what it means to look at paintings and watch plays and read prose and see or hear an individual sitter, character, or narrator within those artistic forms *as a melancholic*. Here visual recognition, clinical observation, medical diagnosis, scholarly interpretation, and playgoing spectatorship all dovetail. They are the means through which someone shows up as melancholic for an onlooker, in an artwork, or simply in a person who stands before us. Wittgenstein, in a handy term of art, called this experience "seeing as," or *aspectual seeing*: "I contemplate a face, and then suddenly notice its likeness to another face. I see that it has not changed; and yet I see it differently. I call this experience 'noticing an aspect'. [. . .] the flashing of an aspect on us seems half visual experience, half thought. [. . .]"[3] This is true not only for Wittgenstein's celebrated optical illusions and duck-rabbits but also in everyday experiences of looking at bodies in motion, and seeing the play of emotional life across the given structure of the body:

> Might I not [. . .] have a purely visual concept of a hesitant posture, or a timid face? Such a concept would be comparable with "major" and "minor" which certainly have emotional value, but can also be used purely to describe a perceived structure.[4]

Such splitting of the perceptual difference between a formal arrangement (anatomy) and an expressive modality (posture) may at first glance seem merely recondite. Yet it can also help us to reactivate the mystery of how we perceive affect in others; Wittgenstein's musical example is particularly important in its insistence upon the coextension and simultaneity of structural expression and cultural meaning. Major and minor intervals are susceptible to an emotionally "colorblind" accounting of the semitones that separate them; they can be structurally accounted for without recourse to any theory of affect or cultural history of their associations with a particular repertoire of moods. Yet Western subjects brought up with an exposure to music created in the wake of a specific, culturally mediated emotional encoding of "major" and "minor" can, and do, detect the relative elevation or depression of "mood" which these forms express, and they can do so without being able to articulate why on a structurally or mathematically precise level. The discovery that a body before us is a melancholic body is, in both visual art and on stage, very frequently a matter of an instantaneous perception of an affect, even if that affect is, at the very same moment, susceptible to a rigorously historical unpacking as culturally inflected convention. As Dryden put it in his "Parallel Betwixt Painting and Poetry": "I must say this to the advantage of painting, even above tragedy, that what this last represents in the space of many hours, the former shows us in one moment. The action, the passion, and the manners of so many persons as are contained in a picture are to be discerned at once, in the twinkling of an eye."[5] Melancholy, for all its centuries of mysterious allure and dense philosophical back story, can also show up for us in a single, immanently expressive gesture: a downcast figure sitting with its head in its hands.

When did such melancholy posturing begin? In the second book of *A Tracte Concerning the Artes of Curious Paintinge* (1584/1598), subtitled "Of the Actions, Gestures, Situation, Decorum, Motion, Spirit and Grace of Pictures," Milanese painter and art theorist Paolo Giovanni Lomazzo dates with enviable precision the first appearance upon the Earth of this emotion, and the characteristic motions that go with it:

> Having generally treated of all the motions, unto what kinde of bodies they do particularly appertaine, and howe they may accidentally befall all sortes of men; it remaineth that I touch each of them severally, beginning

> with MELANCHOLIE: whose motions are pensive, sorrowful, and heavie: and are to be expressed in the picture of Adam and Eve immediately upon their fall, making them with declined countenances, and eies fixed on the earth, bowing the head, with one elbow resting upon the knee, & the hand under the cheeke, sitting in some convenient place, as under a shadowie tree, betweene the rockes, or in some cave [. . .][6]

Eden was the first site/sight of melancholy assemblage. Staging postlapsarian Paradise as the exemplary melancholy scene while providing a direct schematic of melancholy passion as posture, Lomazzo assembles the bodies of Adam and Eve into temperamental shapes, in the process arranging their relations to each other and to the affordances of their surrounding environment. In particular, the action of propping grafts the affective state of sorrowful dejection and the cognitive activity of philosophical reflection onto the sturdy support of a body in space. The posed body acts as an armature upon which to naturalize melancholy as an *attitude*, a term that suggests both an exterior and an interior stance toward the world, both a physical position and a corresponding mental mindset. The natural settings of rocks, trees, and caves anchor these posed figures but foil them too, with their mixture of shade and illumination suggesting both enclosure and exposure. Wedged between the rocks, Lomazzo's Adam and Eve stage humanity as a break or interruption within nature, only partially successful in their search for a place within it. These melancholics are situated in and thinking upon a world which is not yet, and may never again be, their home. Instead, their eyes are fixed upon the earth that will become their grave.

Far from an index of exceptional excellence or personal uniqueness, here melancholy is the most universal of affective positions, our common inheritance. Lomazzo's text announces a basic protocol for melancholy representation already in wide circulation: figures adopting a highly specific, conventional posture in an isolated, natural setting. The stability of this protocol across centuries of Western artworks begs certain questions about the nature of melancholy and, indeed, of what we explain when we invoke a word like "convention" to describe the system of reference through which a body becomes a sign. Cultural materialism would have us historicize such conventions by demonstrating that particular examples of them are subject to location-specific political and economic forces, but in the process, we lose what makes conventions stable across distances in time and territory or, in Deleuze and Guattari's terms, what lends consistency to an assemblage. In this chapter, I would like to consider several exemplary points in the history of melancholy representation—spaced centuries apart from one another—in order to draw out what sort of consistency bodily posture might provide for melancholy assemblages: Albrecht Dürer's *Melencolia I* (1514); Isaac Oliver's *Edward Herbert, First Baron Herbert of Cherbury* (1614); and Bas Jan Ader's *I'm Too Sad to Tell You* (1970). In the process, I hope to pose the following questions: To what extent does the pictorial convention of melancholy posture produce the viewer's experience of affective recognition in each image? Might there be other orders of experience or explanatory rubrics (psychoanalytic, developmental, and evolutionary-psychological) that inflect these historically specific points of tension between personal expression and general convention? What kind of circuits of analogical relationship between figure and ground do melancholy portraits set in place and set in motion? What sort of self-relation does the particular posture of propping transmit, and how does the haptic circuit of self-support and self-touch implicit in propping complicate the visual register?

As Lomazzo's example demonstrates, melancholy's instantaneous legibility as posture offers an index of the sinful collectivity of generic humanity, but it is also imagined as the disappointing consequence of the first act of frustrated "self-fashioning," the sorrowful outcome of a search for knowledge which brings death, and the knowledge of death, into the world. Melancholy affords a particular stance of contemplative reflection upon creation and yet also expresses a precognitive state of dependent embodiment that Walter Benjamin, in *On the Origin of German Tragic Drama*, termed *the creaturely*.[7] "Creatureliness" is the ambivalent estate given to the created body and inherited from Adam—a state of miserable separation from God that also functions as a priming condition for humanity's intellectual flight outward toward divine insight and knowledge:

> If melancholy emerges from the depths of the creaturely realm to which the speculative thought of the age felt itself bound by the bonds of the church itself, then this explained its omnipotence. In fact it is the most genuinely creaturely of the contemplative impulses, and it has always been noticed that its power need be no less in the gaze of a dog than in the attitude of a pensive genius.[8]

As created beings, the dog and the pensive genius both share "the misery of the creaturely estate." Here Benjamin clearly alludes to the dog and angel of Albrecht Dürer's *Melencolia I* (1514), the most celebrated emblem of melancholy and, in several senses, the "first" image of melancholy posture.

That image arrives as a uniquely saturated site of relational networks. If the mournful companionship of angel and dog suggests that the condition of passionate creatureliness is shared across the species barrier, the figure/ground connections between that shared world of organisms and the obdurate block digs still deeper. The stone flags a level of telluric material status located still further down the ontological chain and indexes a melancholy of the dead environment which Scott Wilson has christened "melancology."[9] Black bile cannot flow through the stones of the earth and yet, if Earth is the inmost element of melancholy itself, then the Block incarnates a thingly melancholy which, however geometrically wrought, is precognitive, even pre-affective: The mute and still lump of the block constitutes a primal instance of earth as the elemental bearer of the melancholy of matter itself.

Regularly invoked when discussing artworks that are alternately said to be metaphorical, allegorical, iconographic, or iconological, *Melencolia I* is now overdetermined by its own canonical centrality to the art-historical theorization of the visual image as a bearer of implied meaning. Panofsky's declaration is definitive: "The influence of Dürer's *Melencolia I*—the first representation in which the concept of melancholy was transplanted from the plane of scientific and pseudo-scientific folklore to the level of art—extended all over the European continent and lasted for more than three centuries."[10] In subsequent engravings, frontispieces, and paintings, the melancholic is repeatedly portrayed as seated, holding, or somehow propping up the head, which lolls in an attitude of weariness and despair in a manner that reiterates the attitude of Dürer's angel. As attested by countless exhibition catalogues and anthologies of melancholy portraiture, this posture became a visual commonplace. A representative selection of examples would include Virgil Solis' engravings of "Melancholicus" in *The Four Temperaments* (circa 1530), H. S. Beham's *Melancholy* (1539), Matthias Gerung's *Melancholy* (1558), Domenico Feti's *Melancholy* (1620), Christoph le Blon's frontispiece to the 1628 edition of Burton's *Anatomy of Melancholy* (in which both Democritus Abderitis and the allegorical figure

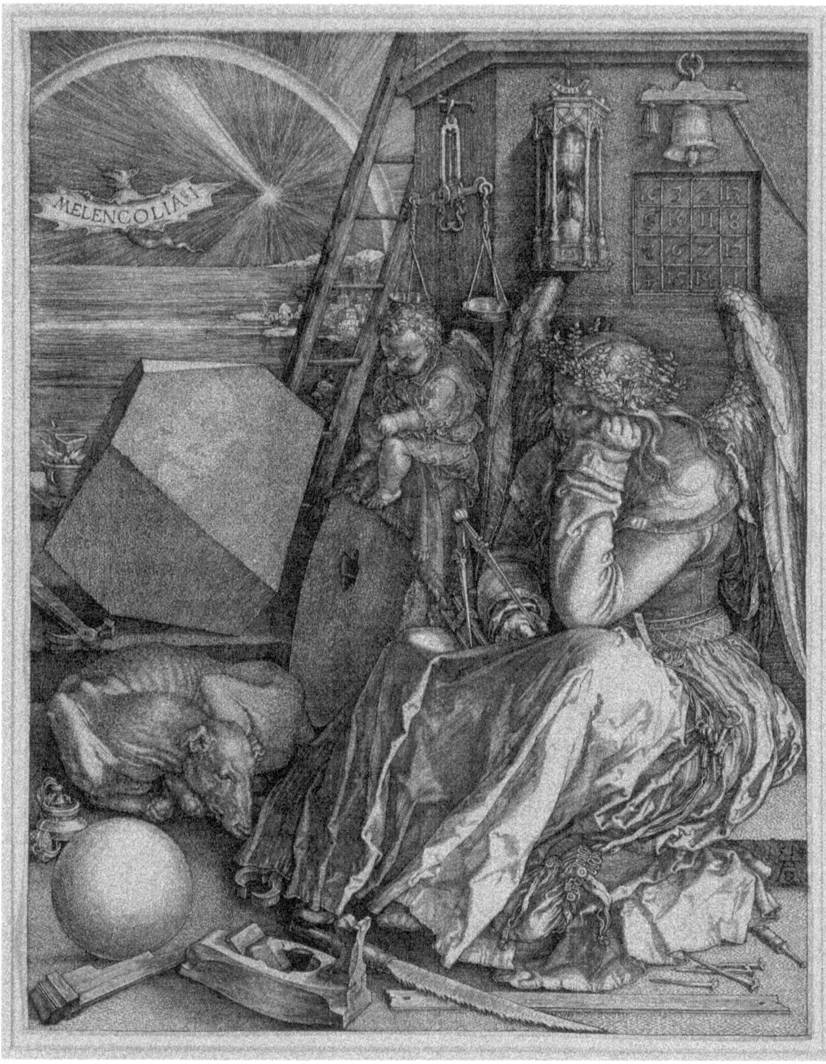

Figure 3. Albrecht Dürer, *Melencolia I*, 1514. (Réunion des Musées Nationaux / Art Resource, NY)

of "Hypochondriacus" adopt the posture), and Salvator Rosa's painting and engraving *Democritus in Meditation* (1650).[11] Among this constellation, I shall linger upon Isaac Oliver's *Edward Herbert, First Baron Herbert of Cherbury* (1614) as a particularly revealing expression of both the consistency and the elasticity of this mode, which persists long past early modernity. Among many examples from eighteenth-century art that self-consciously cite this keynote of melancholic self-presentation, there is a highly explicit return to Oliver's portrait in Joseph Wright of Derby's *Portrait of Sir Brooke Boothby* (1781).[12] Arguably, this pictorial tradition reaches its late contemporary apotheosis in Rodin's *Le Penseur* (1881), in which the melancholy posture becomes elided with the activity of thought itself.[13] Across this broad temporal and geographical range, the propping posture remains a curiously static constant in the tidal changes of figurative art.[14]

Melancholy and/as Posture from Dürer to Oliver

Despite its instantaneous familiarity, there is remarkably little reflection upon the meaning of this posture itself within the abundant criticism of Dürer's image and its aftermath. The self-propping stance expresses melancholy so baldly, so directly, so apparently, that in a sense it hardly requires explanation, only recognition. In speaking of Dürer's shift from early drawings to the final version of *Melencolia I*, Klibansky, Saxl, and Panofsky note "it is unnecessary to add that the upright figure in the preliminary study has been deliberately changed to a drooping one."[15] It is not necessary to mention this change not only because a comparison of the two drawings immediately conveys this shift but also because this fact itself is so obvious, so legible, so utterly conventional, as to go without saying. Dürer's image intervened upon the convention, reduced it, and sharpened its postural focus. Indeed, if we compare Alberti's description of melancholy posture as "drooping" in 1435 with Lomazzo's insistence in 1584 upon the specificity of the propping up of the head, one is tempted to wager that the shift from one to the other turns upon the triumphant viral circulation of Dürer's image as the preeminent model for subsequent melancholy representation.

This is not the same as the claim that Dürer was in some sense "first." From the proliferation of anonymous and unattributed medieval images of the four temperaments in English and German sources that predate him, we know that the propping of the head was already a gesture associated with emotional expression, though the fact that it is sometimes attributed not to melancholics but to phlegmatics in the medieval images suggests that it had not yet taken on its decisive association with melancholy before the early sixteenth century.[16] Regardless of who was really "first" within the archive of melancholy images, the upsurge in representational frequency of this posture after Dürer seems to demonstrate Panofsky's claim about the paradigmatic influence of *Melencolia I*.

But Dürer might not be basing this posture solely upon other artworks and theoretical rubrics that he might have seen. The source may lay closer to home. Some scholars have speculated that the posture of Dürer's angel might be based upon his bride Agnes Frey; as biographer Jane Campbell Hutchison describes her appearance in an early drawing, "she appears heavy-eyed, slightly sullen, and very young, and is seen in much the same posture that Dürer was later to use for the personification of Melancholy."[17] That said, some of Dürer's critics have also detected, beneath the resemblance to Agnes, a deeper current of connection between the facial expression of the angel and Dürer's mother Barbara.[18] Figured thus, the stage is set for a certain sort of overfamiliar psychoanalytic reading in which the erotic object choice of Agnes barely conceals an Oedipal longing for Barbara, with the melancholy angel ultimately indexing the original lost object of the mother's body whose canceled source of nourishment and plenitude thus produces a symptomatic affective mixture of sorrow and protest. While the overfamiliar is not therefore the incorrect, I wish to suspend the force of this identification at the level of *roman-a-clé* in order to preserve a broader range of relational forces which melancholy posture makes available: relations between the parts of the body to one another, relations between the body and surrounding objects, and relations between the viewer and the total scene presented before them. A too-brisk tunneling "inward" to the psyche of Dürer in search of the fixity of an identifiable "lost object" would flatten the reach of the melancholy image in the process, annexing early modern melancholy to Freudian accounts of "melancholia."[19] Though we shall return to the question of the relationship between the mother and melancholy at the close of this chapter, it must be said that the narrowly personal identification of Dürer's melancholy angel with the artist's mother cannot answer the broader questions about the pan-European consistency and conventionality of this posture.

Those who followed Dürer's example absorbed not the "who" but the "what": an arrangement of limbs in space.

Bracketing this unresolvably speculative question of personal psychobiographical origins and granting the broad range of possible art-historical precursors, it is clear that from Dürer's sources to his own engraving in 1514 and on to drawings and sculptures executed four hundred years later, more or less similar postures persist across differences of time and place and continue to code within the present as "melancholy." If, as James Elkins notes, "before it became popular in the humanities, semiology was a branch of medicine, concerned with the ways in which the body could be read," humoralism constituted one of the most powerfully influential forms of medical semiology; in order to function diagnostically, the scheme needed to become visible within and upon the structure of the body as an expressive pattern of disease.[20] Within that practice of reading from outer to inner and back again, Dürer's assemblage of the angelic body and its surroundings visually insists upon the acute centrality of this posture within the frame of the image itself: Looking at the image, our eyes are pulled to the darkened face of the angel cradling her head. By extension, that postural image both inherits and solidifies an entire pictorial tradition, lending consistency to the total assemblage of melancholy images constellated around Dürer's expressive core.

Having thus sketched the overdetermined question of the origins of melancholy posture, I now wish to examine a particularly telling example of its social circulation and re-presentation in early modern England. Telling, for my purposes, insofar as the propping, pensive gesture of Dürer's angel, and the creaturely burden it transmits, arguably provides a catalytic structure for the English phenomenon of fashionably "melancholic" society portraiture, an outcome ironically at odds with the economic imperative of such commissions to render the specificity of particular persons. Since in these paintings concrete individuals are conforming themselves to the terms set by an angelic archetype, we witness in the citation of this posture a crucial example of the expressive labor of conventionality itself as a means through which to attenuate and soften subjectivity rather than as a means to sharpen and distinguish it. As art historian Richard Wendorf describes in his account of early modern English portraiture, the imperatives of convention tend to upstage the representation of individual character:

> The combination of allegorical motifs with the static poses and inscribed legends of many Tudor portraits (such as Gheerhaert's Ditchley portrait of Queen Elizabeth) produced representations that we would call overdetermined; these paintings are so saturated with iconic devices that they are unable to suggest the character of the sitter except in hierarchic and formulaic terms. Even when painters such as Eworth attempted to combine natural (that is, comparatively realistic) descriptions with allegorical motifs—as in his unusual portrait of Joseph Luttrell—the two forms of representation remained strangely divorced from each other, carefully separated in terms of proportion and by the spatial structure of the work.[21]

Against such a backdrop, notions of artistic identity, or indeed of art as first and foremost a matter of personal expression and formal, medium-specific innovation (to risk a modernist cliché), tend to expand and contract in response to market demands for a given, conventional representational mode. Within these highly stylized images, the conventionality of the melancholic posture inherited from Dürer constitutes melancholy portraits as particularly vivid examples of this "strange divorce" between subjectivity and social ground. With this dynamic in mind, I wish to examine the implications of posture's consistency (and its limits and edges) within this tradition through a reading of a single—late, exemplary—work: Isaac Oliver's remarkable portrait of Edward Herbert, first Lord Herbert of Cherbury.

The work is both "late" and "exemplary" in its revival and reuse of components that had flourished on the continent far earlier. Isaac Oliver was the pupil of the English master miniature painter Nicholas Hilliard, and part of what makes Oliver's portrait of Herbert particularly striking is its odd return to the habits of his master; it occupies an awkward halfway point between each artist's signature style. Though it was Hilliard who taught Oliver the art of painting miniatures, Oliver's distinctive naturalism and adventurous use of shade and perspective contrast clearly with the brighter, flatter, and more metaphoric works of Hilliard. Mary Edmond, in her scholarly history of the two artists, draws a sharp distinction between the master and the pupil, effectively defining the succession from Hilliard to Oliver as a shift from the iconographic culture of emblematic painting to "realism" full stop: "Oliver's miniature exemplifies [a] realism . . . in total contrast to the stylistic imagery of his former master."[22] Roy Strong in *The English Renaissance Miniature* reads Hilliard and Oliver's connection differently, defining Oliver not so much in terms of "realism" broadly understood but as a continental Mannerist out of his element and ahead of his time in the comparative backwater of Jacobean London.[23]

The passage from Hilliard to Oliver would seem to define a transitional boundary line between two stances within painting, between the emblematic and the real, the coded sign and the physical body, and thus as part of a longer historical arc within portraiture as a practice in transition "from romance to realism."[24] But as many art historians have been quick to point out, this Whiggish narrative of progress from the iconographic to the natural is too linear to account for the complex negotiations between past and present that we discover in particular portraits. Marcia Pointon's description of the epistemological project of eighteenth-century portraiture in *Hanging the Head* informs my own understanding of early modern portraits. In Pointon's words, portraits intrude into the world they depict, for "they open onto a politics of representation in which the human subject is not a separate entity from the portrait depiction of him or her, but part of a process through which knowledge is claimed and the social and physical environment is shaped."[25] The conventionality of melancholy representation offers us a chance to see this shaping of social knowledge take place within and upon the body itself: In keeping with Alberti's and Lomazzo's treatises, the posed body is positioned into a temperamentally legible shape which allegedly expresses the humoral state within. And yet, in making such legibility possible, portraits are, as Richard Wendorf puts it, "necessarily governed (or at least limited) by the pressure of convention."[26]

This may have more to do with what it is like to be within a body than what it is like to pose for a portrait. Whether standing before a canvas or walking down the street, the rendering of what is incorrigibly particular into socially intelligible terms is simply basic to the very experience of how we stake ourselves before others. Sensing this, Pointon's formulation is apt: "At its most abstract, portraiture is a question of the relationship between the self *as* art and the self *in* art."[27] It is precisely this disparity between the reality of Herbert's own persona and its temperamental stylization that the dispute between naturalism and metaphor within Oliver's portrait makes available to its viewers today.

This pictorial split translates a conceptual split within melancholy itself. Because melancholy as an illness hovers between physical and psychological definitions, the activity of reading melancholy portraiture requires an uneasy shift between the clinical observation and diagnosis of visible symptoms and an interpretive decoding of the symbolic correspondence between the melancholic body and its emblematic surroundings. As in Dürer's foundational example, so here the viewer must shuttle back and forth between "Galenic" indicators (dark complexions, bent bodies, and thinness of physique) and "Aristotelian" indicators (iconographic signifiers of excellence, inspiration, power, and extremity). It is particularly telling that Oliver's melancholy portrait should straddle these particular representational models, while not sitting comfortably in either of them. Emblematic painting provides the viewer with objects, signs, and surroundings which are meant to function as analogues to the psychological interiority of the melancholic. "Naturalist" paintings of melancholics simply present the body in a physical posture or attitude which flatly communicates melancholy as an affliction legible in terms of the body alone, without a particular reference toward inner life or mental suffering. The former mode codifies the melancholic body into a scholarly text, ripe for its own Ficinian/Aristotelian contemplation; the latter mode models the individual as a suffering body, prepared for entry into a Hippocratic/Galenic physician's case history.

But by the time of Hilliard and Oliver's miniatures, the depiction of melancholy had definitively shifted from a pathological exposure of the miseries of illness into a fashionably softened idealization. In order to become a suitable attitude for the commission of a portrait, melancholy had a strong economic incentive to segue from a condition of painful humoral imbalance toward one of mournful, contemplative genius. No one was going to pay for a painting that made them look flatulent and miserly, though they might well pay to look amorous and brilliant. But across this cultural shift, the melancholy body remained curiously consistent because its posture, that is to say, the structural arrangement of the creaturely body, "its attitude," remained fixed in the self-propping pose.

With this warning about "stiffness" in mind, what is so striking about Oliver's portrait (Plate 1), is its reactionary, regressive character, its conscious aping of his predecessor Hilliard: An indistinct forest background is used to largely evade any firm perspective, its colors lack the subtlety of Oliver's characteristic work and instead are "lively" in the Hilliard manner and, most importantly for my purposes, Oliver's cabinet miniature incorporates emblematic content in the depiction of the shield. Sounding almost disappointed at the way that this painting, probably executed in 1613 or 1614, "lurches back to the 1590s" tradition of chivalric tilt portraiture, Sir Roy Strong is blunt in his assessment of what is going on: "Oliver gave his client what he wanted, a romantic Hilliardesque vision."[28]

It is tempting to see in this portrait a balancing act between Herbert the philosopher and Herbert the gallant young man, but we ought to resist the tendency to allow Herbert's later achievements in philosophy to be retroactively fixed upon this portrait, and to confer upon it an anachronistic *gravitas*. The Edward Herbert of 1613 was a man-at-arms first and foremost, and the portrait could convincingly be read as not melancholy at all but simply as a representation of confident masculine display in the manner of Nicholas Hilliard's portrait of George Clifford, Third Earl of Cumberland in "tournament fancy dress" painted in 1590 (Plate 2), the year he succeeded to the office of the queen's champion in the tilts.[29]

Herbert's embodiment of a similar chivalric ideal and performance of a quaintly archaic virility staged by the side of this poetic brook, with his lance and horse ready to hand in the background, recalls the function miniatures played in a marriage market, with courtships frequently conducted on the basis of the exchange and public wearing of miniatures. While Herbert himself was married at the time of this painting (to his cousin Mary, with whom he had fathered four children), his painted counterfeit nevertheless went into an erotically charged circulation, as the following account of Herbert's miniature from his autobiography indicates:

> There was a Lady also wife to Sir John Aeres, Knight who finding some meanes to get a Copy did afterwards get it contracted in to a litle forme by Isaac the Painter according to his manner and afterwards, caused it to bee set in gold and Enamiled and soe wore it about Her neck soe lowe that shee yet hid it under her brests, which I conceive coming afterwards to the knowledge of Sir John Aeres gave him cause of Jealousie . . .[30]

The result of the discovery of the portrait was a gruesome roundelay of violence which ultimately led to both parties being called before the law, with Herbert's eventual vindication against the charge of whoring Sir John Aeres' wife belied by the innuendo of his text. Herbert's brief reference to "the painter Isaac" creating a portrait miniature "according to his manner" to be worn by a young lady is tantalizing but unsubstantiated by any existing miniature of Herbert fitting that size and description; the possibility that the portrait referred to is the one under discussion today founders on the rocks of that portraits' unusually large dimensions (7½ by 9 inches). If anything is clear from Herbert's autobiography, it is that during the time the cabinet miniature was painted he was anything but a melancholic and, indeed, his glib catalogue of duels, assaults, and military adventure confirms his passing remark that: "in my owne Nature I was ever Chollerique."[31] This self-understanding is also immediately visible within the image. At the level of emblematic meaning, the shield pulls away from melancholy signification—not in order to reject humoral reading but rather to reorient that reading in the direction of a different humor altogether: choler. The heat of the flaming heart suggests not simply romantic ardor but the characteristic fiery temper of a choleric. It makes the location near the spring within the earth correspondingly therapeutic: if this warrior suffers from a flaming excess of inner passion, then the grounding of the composition at the source of the cooling waters of the spring offers an antidote.

If the sitter's location references the conventional locale of melancholics by suggesting a hermetic retreat from society, that setting could just as easily refer to Herbert's dueling, as duels typically occurred outside the reach of authorities which sought to quell the practice. Indeed, Herbert's description of his preparations for a duel with Lord Walden, Eldest Son to the Earl of Suffolk, outside the city walls ("my self resolving in the mean time to rest under a fair Oak; after [. . .] tying my horse by the Bridle unto another tree"), describes rather exactly the scenario of Oliver's miniature.[32] And yet, despite these counterexamples that suggest chivalric contexts for the portrait's setting or choleric traits in the portrait's sitter, the portrait remains "melancholy."

That it does so is in large part due to posture. Oliver's Herbert lies upon the ground, his head resting on his hand. In a sense, Oliver's Herbert is the final evolution of two earlier paintings which exemplify the pictorial conventions of melancholy: Nicholas Hilliard's *The Young Man among Roses* (Plate 3)[33] and Oliver's own earlier *Young Man Reclining Against a Tree* (Plate 4). Sir Roy Strong provides an intriguing gloss of these shared locations and conventional postures: "As the century draws to its close love melancholy is depicted. Oliver's gallant sits in a 'dump' or a 'muse' with

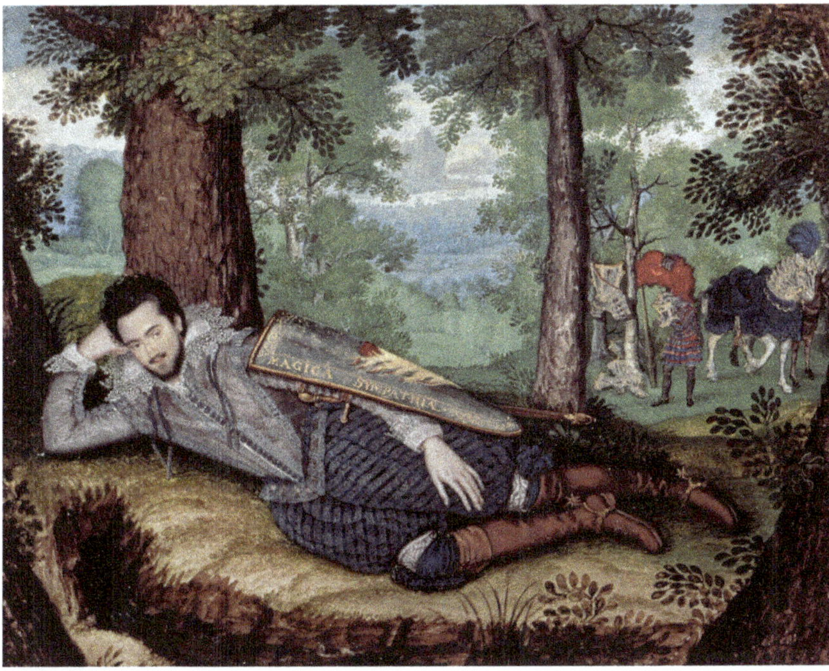

Plate 1. Isaac Oliver, *Edward Herbert, First Baron Herbert of Cherbury*, 1610–14. (Private Collection / The Bridgeman Art Library)

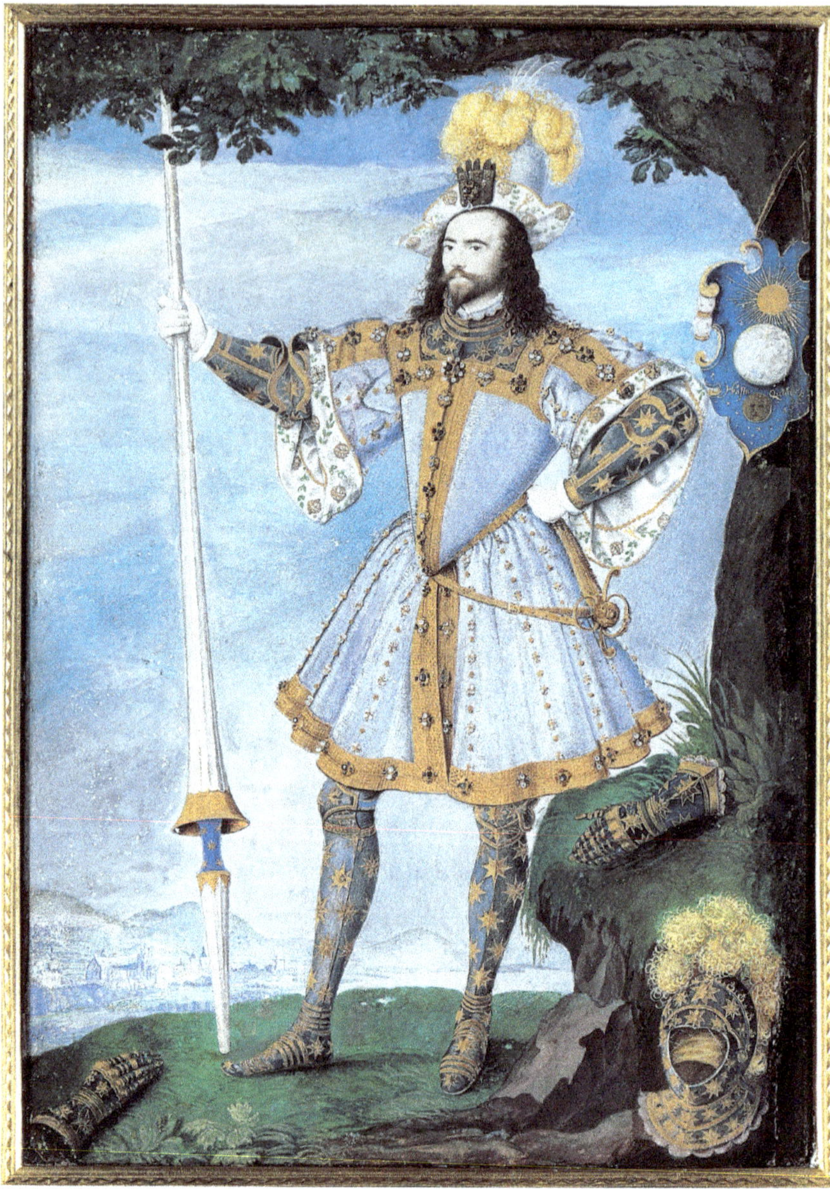

Plate 2. Nicholas Hilliard, *George Clifford, Third Earl of Cumberland*, 1585–89.
(DeA Picture Library / Art Resource, NY)

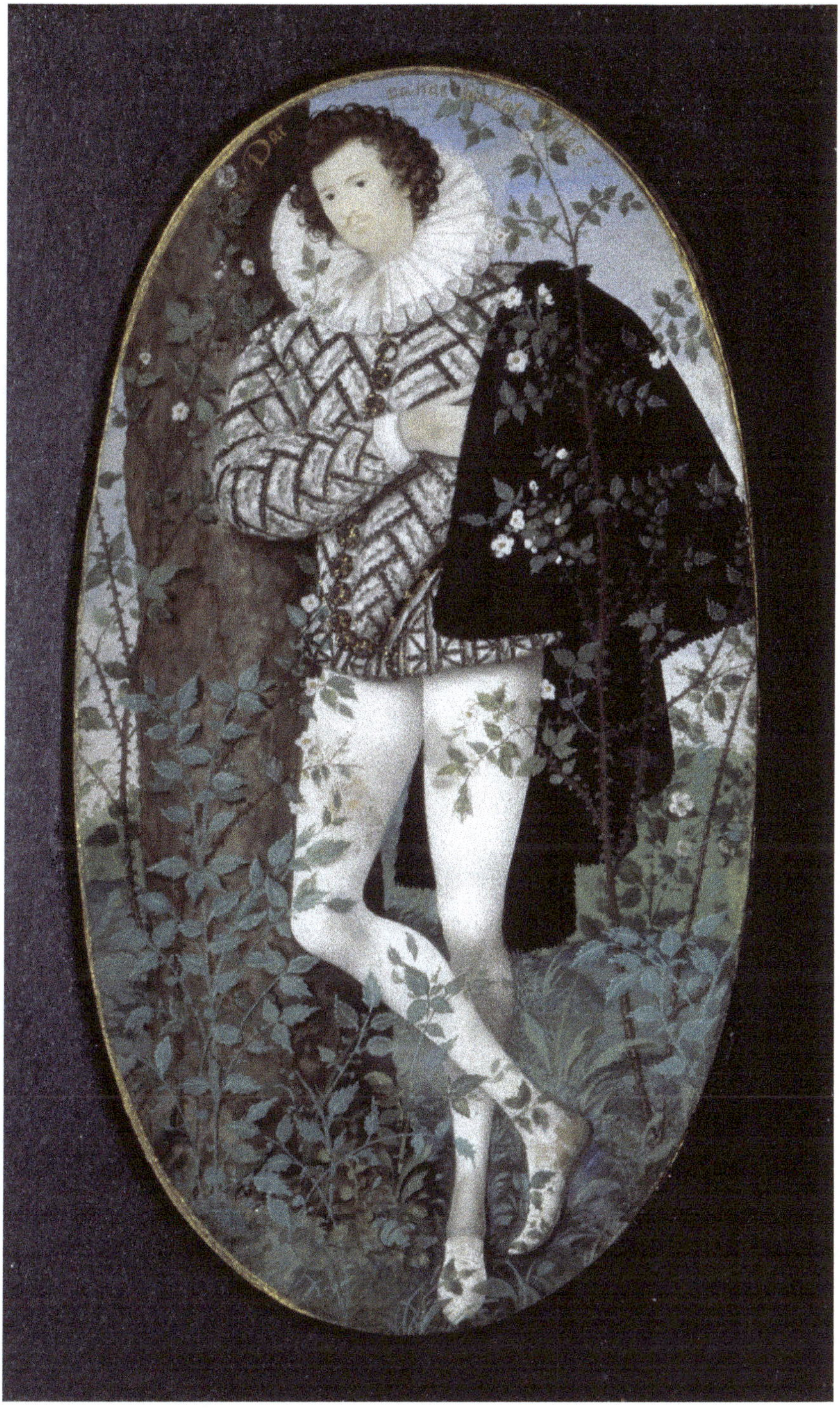

Plate 3. Nicholas Hilliard, *The Young Man among Roses,* 1587. (V&A Images, London / Art Resource, NY)

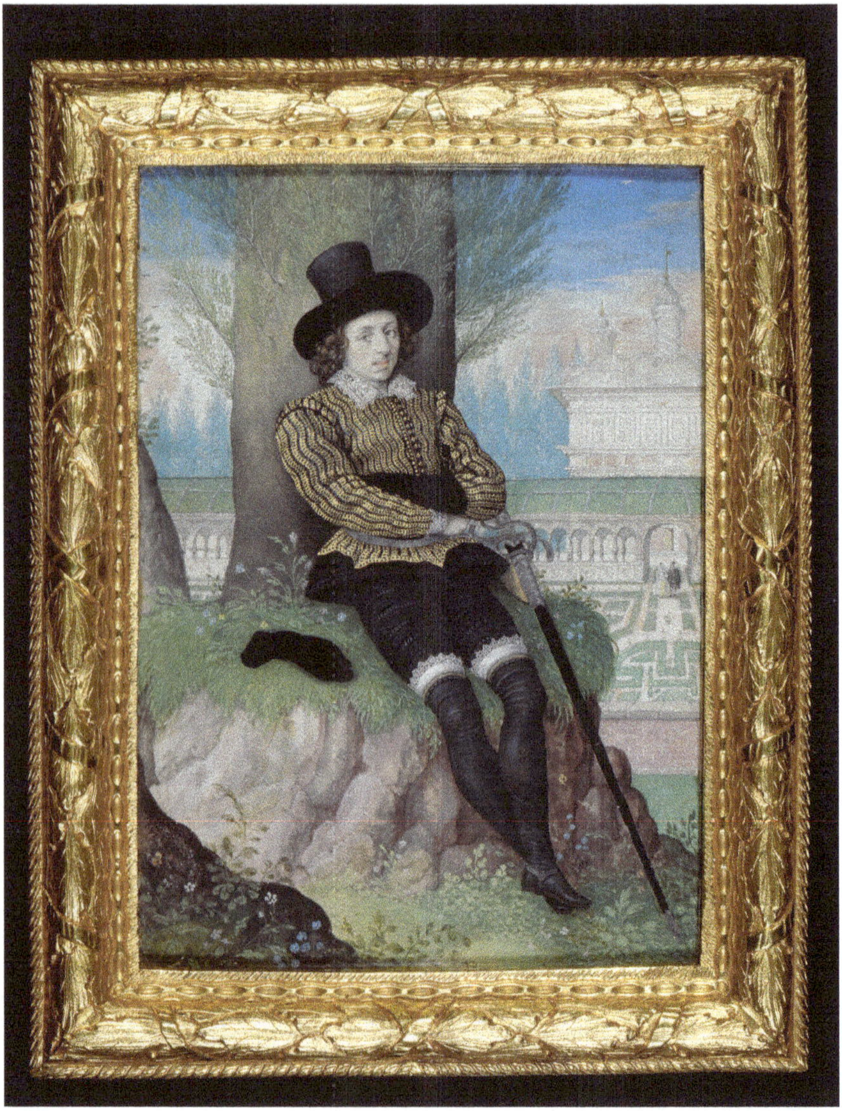

Plate 4. Isaac Oliver, *Young Man Reclining Against a Tree*, 1590–95. (The Royal Collection © 2011, Her Majesty Queen Elizabeth II)

crossed arms and floppy black hat beneath a tree."[34] Resting against natural forms in poses which suggestively braid together isolation, desolation, and contemplation, Strong's equivocation about how to describe the locations of these melancholy miniature portraits ("dump" or "muse"?) perfectly sums up the melancholic figure's precarious positioning between attitudes of depressive suffering and inward illumination.

Encountering Oliver's Edward Herbert in the wake of these two portraits, it is almost as if we are witnessing the same melancholic young man, despairing of company, finally resting upon the ground at the base of the tree, at last holding his drooping head in his hands. This is not surprising from a self-consciously "late" expression of this tradition, for the regressive slump from erect and self-supporting uprightness down toward this original gesture of propping—and its accompanying telluric communion with the ground—demonstrates the lingering magnetism of Dürer's founding example. Though Oliver's subject lies down while Dürer's sits, the return to propping is the fundamental gestural bond between the two bodies, a "clicking into place" that indexes the expressive consistency of the melancholic body as a postural assemblage over time. From within the grasp of the conventional tradition it reinforces, the body assembled into just this form seems uniquely expressive of melancholy, converting the body into an instantly legible ideogram of affect.

Walter Benjamin on the "Idea" as Configuration

Might the body constitute not just an ideogram but an Idea? It is here that Walter Benjamin's definition of an "Idea" provides a helpful model for an interpretation of the melancholy posture as a directly expressive structural form. In his essay "The Ruins of Walter Benjamin," Charles Rosen outlines the specific German intellectual history behind Benjamin's notion of "Idea," drawing a sharp and critically useful distinction between mere concepts, which have a propositional form, and "Ideas" proper, which have a specifically nonpropositional status. Rosen argues that Benjamin's sense of the term derives from Novalis, whose notebooks draw the distinction in the following manner:

> The [artwork] as such contains no defined result—it is not the picture or the factual reality of a proposition. It is the graphic carrying-out—the

> realization—of an Idea. But an Idea cannot be seized by a proposition. An Idea is an infinite series of propositions—an irrational quantity—untranslatable (musical)—incommensurable. [. . .] What can be set down, however, is the law of its development.[35]

So Ideas are nonpropositional sets of propositions. But, as Rosen points out, Benjamin uses two other metaphors—configuration and constellation.

> The representation of Ideas takes place through the medium of empirical reality. For Ideas are not represented in themselves, but solely and exclusively as an arrangement of real, concrete elements in the concepts. And indeed as the configuration of these elements . . . . The staff of concepts which serves as the representation of an Idea realizes it as such a configuration . . . Ideas are to things as constellations are to stars. This means, in the first place, that they are neither their concepts nor their laws.[36]

How can we understand Benjamin's relation between the Idea and its constituent "things"? One useful comparative model might be Althusser's notion of expressive causality—a structure is said to be the "cause" of its effects and its effects are taken to be the total body of its component parts—but that structure is not separable from the contingent arrangement which expresses it: "The structure, which is merely a specific combination of its peculiar elements, is nothing outside of its effects."[37] Far from being simply one of the elements within a broader constellation, the melancholy body, legible in its conventional posture and setting, is itself such an "Idea." The melancholy body is a constellation or configuration in the specific sense that Benjamin develops it in *On the Origin of German Tragic Drama*: It does not have a propositional character; it does not yield a single, united, literal "meaning," but it is recognizable; it communicates; and it generates an indefinite, potentially infinite series of assemblages that structurally express its Idea. This intransigently nonpropositional character ensures that the promise of a corresponding interiority which will speak its truth to us, a "concept" or motto which is being stated by the picture of the body, will always be frustrated, even as the ease with which we recognize and "see" melancholy-as-posture demonstrates to us that we already know what a given assemblage of the body is saying. Melancholy so understood is not the interior psychological state that a given selection of symbolic objects expresses, but the fact of the body itself. That is to say, melancholy *is* posture.

And yet, of course, Hilliard's decision to represent the decidedly unmelancholic Edward Herbert in this conventional mode is also something of an imposture too. If the chivalric masculinity of the tilt portrait offers one mode through which Herbert can here stage himself, perhaps melancholy here offers a flattering aura of sensitivity as its complement. Lacking both parodic critical thrust and the interest of pathological exemplarity in depicting madness, in Oliver's miniature and in works like it melancholy is visually retooled and circulates not as a sign of humoral discomfort or crippling psychosis but as an entirely attractive cultural location of judgment and taste. To be melancholic is not to suffer from one's very interiority but to perform with one's body a surface attitude, to exhibit a mode of being that is inevitably subject to iteration and consumption. For Herbert-as-perceived-by-Oliver, it is to enter into a playful imposture and—along with Jaques and the lovers, courtiers, lawyers, musicians, scholars, and ladies—to join in the fashionable dance.

The Shield as Impresa *of Physical Vulnerability*

To take up and enjoy this surface is tempting, but when we look again at the painting, something sticks out. Can we read this as a painting of and about surfaces without propositional content given the strong counterexample of Herbert's flagrantly overcoded and textualized shield? This shield, which bears an emblem of a flaming heart and has an accompanying tag (the words *Magia Sympathia*) is one of the primary details which stamps Oliver's work as "Hilliardesque," pulling this painting from 1613 backward in time to the 1590s. For while Hilliard's miniatures frequently bear inscriptions or mottos which display his elaborate calligraphic hand, Oliver abandons inscription early in his career (in the process making attribution more difficult) and rarely includes text upon the surface of pictures. While Hilliard frequently incorporates emblematic devices (hands emerging from the sky to shake with the sitter, color-coded pansies meant to represent love-in-idleness), Oliver generally eschews emblematic content. This inclusion of emblematic content on the picture surface is not merely an accident of Hilliard's style but a reflection of market demands. In the context of the English marketplace, the phrase "Magia Sympathia" is typical of what Richard Wendorf terms *iconic portraiture*, an "allegorical representation of character" in which "portraits attempt to augment visual properties with verbal cruxes."[38] But this crux decorates the surface of a shield rather than floating freely upon an otherwise naturalist image.

Hilliard's status as limner to the court required both the painting of miniature portraits and the fashioning of *impresa*, the designs upon the shields used for the tilts, which typically featured allegorical images and short Latin tags or mottos. It is likely that Oliver as his apprentice would have assisted in or at least been familiar with their construction. In *Remaines Concerning Britain*, William Camden offers a striking definition of the *impresa*:

> An *Impress* (as the Italians call it) is a device in Picture with his Motto, or Word, borne by Noble and Learned Personages, to notifie some particular conceit of their own, as Emblems . . . do Propound some general instruction to all . . . There is required in an *Impress* . . . . a correspondency of the picture, which is as the body and the Motto, which as the soul giveth it life.[39]

Camden's lively analogy asserts the necessity of both picture and motto to a fully legible *impresa*, but what is not altogether clear in this correspondence theory is the nature of the dependence between its terms. If bodies require souls to be given life but souls can persist and circulate after the death of the body, it would seem that a subordination of picture to propositional content is at least implied. Following Camden's logic, if the pairing of Noble Personage/Impresa is itself mapped onto the opposition's picture/motto, and body/soul, it is as if the impresa's propositional content were meant to somehow express, preserve, and animate the body it shields. In a sense, tilt portraits such as Hilliard's portrait of the Earl of Cumberland and Oliver's portrait of Herbert—in which the individual body of the chivalric male is depicted accompanied by their corresponding impresa—worry at the ontological status of both the shield and the body. Even as the impresa fixes the family name to a specific propositional content by pairing the impresa with the body it designates, the creaturely contingency of the body is actually drawn out and magnified.

In a development of Hegel's remarks on chivalry in the second volume of his *Aesthetics*, Benjamin connects personal honor, shields, and creatureliness in the following highly abstract and idiosyncratic formulation:

> The dominant role of honour [in Trauerspiel] derives from the creaturely estate of the dramatic character. Honour is, as Hegel defined it, the "extreme embodiment of violability. For the personal self-subsistency for which honour contends does not assert itself as intrepitude on behalf of a communal weal [. . .] On the contrary it contends simply for the recognition and formal inviolability of the individual subject." This abstract inviolability is, however, no more than the strictest inviolability of the physical self. [. . .] The name which, in its own inviolability, claims to bear witness to the apparently abstract inviolability of the person, is in the context of the life of the creature, as opposed to religion, nothing in and for itself; it is only the shield designed to protect man's physical vulnerability.[40]

Though it is unclear from the context of Benjamin's selective quotation, for Hegel, honor embodies violability because it can only be contractually established, that is, it obtains between individuals, and only because of one person's willingness to ascribe honor to another.[41] Hence chivalry can be violated so easily. In the face of this contingent violability, honor asserts a formal inviolability, that is, an individual subject which could not be violated or breached is posited as honor's recuperated ideal, and this ideal is necessarily imagined or projected as an inviolable image of the physical self.[42] For Benjamin, the name shares this quality with that imagined body—it is abstract, inviolable, and meant to stand in for the contingent and threatened human body it designates. This relation between the name and shield—or rather, the name as shield—and a protected, vulnerable physical body beneath that sheltering name recalls Oliver's portrait of Edward Herbert. The inviolability of the name, which its special character as language carries, allows it to both designate a vulnerable, physical body and to evade the threat of violation attendant on such contingent physicality. If Oliver's portrait protects Herbert's body, his physical creatureliness, beneath and behind a literal shield, it also performs a representational work of protection at the emblematic level. The shield furthers the project of family honor by demonstrating individual prowess and affixing the contingency of this individual body to the abstract stability of a shared and legible sign.[43] The ascriptive dependence of chivalric honor is, in this respect, rather like the ascriptive dependence of melancholy itself: It requires ratification to emerge within the social ontology of what can be recognized.

Offering a bid for such ascriptive recognition but destabilizing that bid at the same time, Oliver's painting is transitional in its mode and critical in its effect. While it functions as an impresa by linking the motto ("Magia Sympathia") and image (a flaming heart) to its correspondent family name "Herbert," it also qualifies that functionality in the ways that it stresses the "thingness" of the shield, and thus fails to flatten the impresa into a "pure" emblem. Oliver exploits his medium expertly to bring this about—for the shield is itself tilted away from the viewer and into the picture plane, thus remaining legible enough to transmit its emblematic message yet also marked by the contingency of the real, with its position as an object in space obscuring one half of the shield, and casting a shadow across its surface. Rather than flatly performing its emblematic role, the tilting shield literally tilts away from us, flickering unstably between the emblematicity of the sign and the thingly weight of human equipment.

I wish to suggest that at a second order, this very instability constitutes the representational dynamic of melancholy's appearance for its onlookers, the hook that pulls the viewer into a socially extended network of melancholy assemblage with the body under observation within the image: We are asked to toggle back and forth between an insistence upon the discursive universality of a linguistic sign ("melancholy," a word anyone can learn to use and a posture anyone can learn to fake) and an equal insistence upon the incommensurable specifics of the individual human person ("melancholy," the most intimate experience of self-reflexive solitude, of being-one's-self, which is possible). As with Jaques' simultaneous declaration that all forms of melancholy are socially shareable masks and stereotypes and his claim that somehow his melancholy remains "his own," the tension between Herbert's shield and his body presents melancholy—like Old Hamlet's ghost—in a "questionable shape," stranding him somewhere between type and individual. To use Benjamin's terminology, Herbert is lodged between *convention* and *expression*, comfortably occupying neither term. As with Jaques' acknowledgment and self-distance from "the lover," "the soldier," "the courtier," and others, Herbert here partakes of these referential assignments—but at arm's length. At once exposed to us and shielded from us by a melancholy posture that arrives ready-to-hand after centuries of precedent, we cannot fail to see him as melancholic, yet we do not quite believe him either. This basic oscillation between surface and depth, between language and the body, between Galen and Aristotle, epitomizes the figure of the early modern English melancholic as a relational assemblage, an assemblage-of-assemblages nested within one another.

Faced with the intellectual provocation of this flickering instability in visual experience, we are made aware of our own position as spectator. This characterizes not just the visual experience of looking at a melancholy portrait but also obtains in the experiences of playgoing and textual reception that will make up the remainder of this book. Such an experience "relocates" the phenomenon of melancholy itself, shifting its terrain from the inaccessibly private interiority of a lone sufferer (Jaques alone: a single person model; Edward Herbert alone: a single person model) to the social field in place between the sufferer and the observer(s) who witness that suffering (Jaques-and-Rosalind or Jaques-for-Rosalind: a two person model; Herbert-for-Oliver or Herbert-for-the-viewer: a two-person model). Melancholy is always melancholy for an additional someone who is confronted by it, for whom it shows up as a puzzle to be solved—a problem to remedied—in an act of interpretation or ascription on the part of a second party.[44] A critical focus upon that act of ascription expands the terrain of what the critic of melancholy will be responsible for explaining. Of course, the symptomatic self-scrutiny that characterizes melancholy in its genial form potentially locates this "two-person" dynamic within the psyche of the melancholic himself, with early modern subjects speculating and interpreting the cause of their own melancholy through a kind of analytical self-distance or intra-psychic split crystallized around their subjective experience of themselves as melancholy subjects. At certain points in his soliloquies and melancholic asides, Hamlet-for-Hamlet may very well function as "a two-person model" in the manner I am intending; certainly, Ficino's fantasy of the melancholic sage contemplating himself at the center of the earth epitomizes this image of melancholy as the deepest (in every sense) form of self-knowing:

> [. . .] for the pursuit of the sciences, especially the difficult ones, the soul must draw in upon itself from external things to internal as from the circumference to the center, and while it speculates, it must stay immovably at the very center (as I might say) of man. Now to collect oneself from the circumference to the center, and to be fixed in the center, is above all the property of the Earth itself, to which black bile is analogous.

> Therefore black bile continually incites the soul both to collect itself together into one and to dwell on itself and to contemplate itself. And being analogous to the world's center, it forces the investigation to the center of individual subjects, and it carries one to the contemplation of whatever is highest, since, indeed, it is most congruent with Saturn, the highest of planets. Contemplation itself, in its turn, by a continual recollection and compression, as it were, brings on a nature similar to black bile.[45]

Ficino's portrait of the melancholy child of Saturn buried at the core of the Earth through the sheer focus of their own heroic act of self-reflexive contemplation sets in place an iconic image of melancholy as the proper posture of thought itself. For Ficino, contemplation *is* melancholy and melancholy *is* contemplation. The nobly antisocial isolation of the melancholic from all other human beings is the price of access to a state that is simultaneously the furthest inward and the furthest outward that one can go.

Such feverishly analogical inflations of melancholy with cultural capital and philosophical *virtu* call forth the melancholic sage's comic reflection: the foppish pretender to melancholy eager to bask in the borrowed allure of the initiated. As Lawrence Babb notes: "Clearly in Renaissance Italy melancholy was regarded as the malady of great minds. It is a fair assumption that it was affected by the would-be great."[46] The more esteemed melancholy was as the putative possession of an eminent, influential, and select few, the more tempting it was to attempt to acquire melancholy's associative glamour by taking on the characteristic mien, habits, and mannerisms which made this identity socially legible to others. Just as the circulation of gold produces the haunting possibility of coin clipping and counterfeiting, so too did Ficino's cult of melancholy as the marker of scholarly authenticity, civility, and greatness set in motion a continental vogue for the sorrowful, thoughtful countenance, and with it an attendant insecurity about what lay beneath the melancholy posture. Herbert's portrait—with its casual yet studied engagement with postural convention—suggests the allure of melancholic identification as an accessible means through which to suggest profundity and passion. Yet it also indexes the proximity of those ideals to their debased forms: pretension, foppery, boredom.

Before the stage is set for an impossible attempt to distinguish "real" from "false" melancholy, we ought to admit that the substantial dullness of sitting for one's portrait might induce a state of stupefaction and lassitude perilously indistinguishable from its more illustrious relative. Citing Vasari's recommendation that players and singers be brought in to amuse clientele growing restless during their own portraiture, Harry Berger Jr. argues in *Fictions of the Pose* that we "ought to take seriously the merely practical turn Vasari gives to the term *melancholy*, which he uses to denote not a humoral disposition or dominant character trait but a reaction to the boredom of sitting."[47] This pressurizes the problem of the resemblance between "real" states of reflection and their pretentious imitation from a different direction. As anyone who has smiled for the camera a little too long already knows, moods assert themselves and recede. Riding these waves of emergence, we can go from "real" to "fake" versions of an emotional attitude within ourselves over time. But in noting the difference between the stable image of the "subject of the portrait" and the temporally extended "sitter for the process," we also must grant that there may never have been a singular "state of mind" within the sitter that precisely corresponds to the moment of the image. These implicit fractures of self-difference are part of melancholy's own suspension of action and aspiration toward stasis. While the melancholy subject seems carelessly immersed in a mood that they also reflect upon, the posed body of the melancholy "plays possum" for our view, holding a position for an implied viewer who is invited to recognize the presence of melancholy affect on or in that body. As Berger puts it in his reading of Lotto's *Portrait of A Man (The Dyspeptic)* (1535), "attention is redirected from whatever he may feel to his interest in displaying it."[48]

This representation of melancholic interiority as an intermittently self-present and self-critical subjectivity which takes up its own embodiment and psyche as a problem to be solved leads to an increasingly complex intersection between the historical trajectory of melancholy representation and the abandoned battlefield of critical narratives about the supposed emergence of "modern subjectivity" within the period.[49] The passé nature of such discussion should be taken to mark an impasse rather than a solution, but melancholy subjectivity has perhaps always shown up for itself as already in some sense passé, tethered to a prior cultural moment upon arrival. The citationality of black bile as a Renaissance encounter with an unverifiable but theoretically inherent substance supposedly within one's self models not just the "polychromic" but the self-belated, for in the moment of symptomatic presentation the Renaissance subject sees herself as already in some sense prefigured by "ancient" medicine. What is emergently "modern" about this self-diagnostic positioning is that the shriveling of the substantial basis in favor of an instantly recognizable, eminently falsifiable, pose turns the supposedly "deep" telluric grounding of melancholy into a surface/structure dyad that is legible at a glance.

A Pose Is a Pose Is a Pose

Are we left with a purely discursive reduction of this posture to the history of its dissemination, with Dürer occupying a *faute de mieux* position as the origin? This chapter began with one naïve question: Who was the first melancholic? I now wish to ask another: Bracketing its art-historical success story, why is *this* particular posture associated with both sorrowful, dejected affect and philosophical contemplation? What is it about propping or supporting the head that conjures such breezy, apparently intuitive associations for onlookers? And is there any scandal in the forking paths of those associations toward the conjoined registers of thinking and feeling? When we support our head with our hands, why does this connect us—supposedly—to an emotion of sorrow and despair *and* an action of abstract reflection? The sense of the head as weighty—as a burden—would suggest that it is a problem, in either our intellectual or emotional life (or both at once), that wants to draw support from the stolid fact of the body. Speaking loosely, we tell ourselves that we need to get ourselves grounded, to screw our head back onto our shoulders, and so on. This is a posture that arrests movement and fixes into place a proprioceptive circuit between the surface of the hand and the surface of the head, and it is at the membrane of this recursive feedback loop of self-feeling that I want to speculate—in a loosely phenomenological mode—about a possible antecedent scenario, a haptic memory that might hover at the point of auto-contact.

Having bracketed a too-narrowly psycho-biographical identification of Dürer's angel with Dürer's mother, I now want to return to mothers but of a nonspecific kind. If you think about the way that a newborn baby is held, very often the mother will take especial care to support and cradle the delicate infant head. The head will be supported by the mother's hand and held in place. When we prop up our own heads, we might be triggering a deep bodily memory of such gestures of maternal care and support, effectively citing a moment when we were protected and loved and playing it back to ourselves. But because we are doing this to ourselves, this memory also reminds us in the present moment that the mother is not there to support us, not there to keep us safe, and so there is a kind of necessarily failed moment of self-mothering that is going on in this posture. Because it is an awkward posture to hold for very long, the inevitable break or cut when the propping act of support is given up signals that all such care has to end sometime, and this reminds us again of what we already know: We are on our own in this posture, and for a limited time.

Such speculations are necessarily provisional, but their reach toward a universal developmental scenario provides one possible way out of the falsely discursive premise that there is an art-historical moment of creation *ex nihilo* to which we must assign the name of a particular painterly "author function" ("Dürer") in response to the question of how melancholy posture began. Given the virtual shape-space of the body as a flexible, evolving form capable of a spectrum of possible assemblages of its constituent parts (and in which a possible assemblage is one in which the arm supports the head), any question about origin myths, about the "first image" of a slumping body organized into such a posture, is inherently unverifiable, perhaps infinitely regressive. The dawning of a new convention on canvas or paper implies a prior moment of posing and arrangement within the studio, and that momentary pose implies the possibility of some still prior moment that predates artistic representation itself. Before Agnes Frey slumped at the kitchen table in front of her new husband Albrecht Dürer, and before Barbara Dürer held the delicate head of her son Albrecht, countless unrepresented bodies had already been held, had already slouched and slumped. Therefore, to give Lomazzo's origin myth its due, the answer to the question "who was the first to strike the melancholic posture" is, simply, everyone.

The indistinction of that Edenic "everyone" ripples past the category of humanity itself. The problematic knot of self-care and remembrance summoned up by this posture may not even be specific to human beings if we consider that many primates also adopt this posture, an extension of melancholy across the species barrier consistent with both Benjamin's ascription of melancholy to the gaze of a dog and, more recently, Gail Kern Paster's reflections upon the melancholy of "gibb'd cats."[50] And it may go beyond even the memory of mothering when we consider this image from the groundbreaking social-isolation experiments of Harry Harlow, subtitled "self-destructive behavior in an isolated male rhesus monkey" (Figure 4).[51]

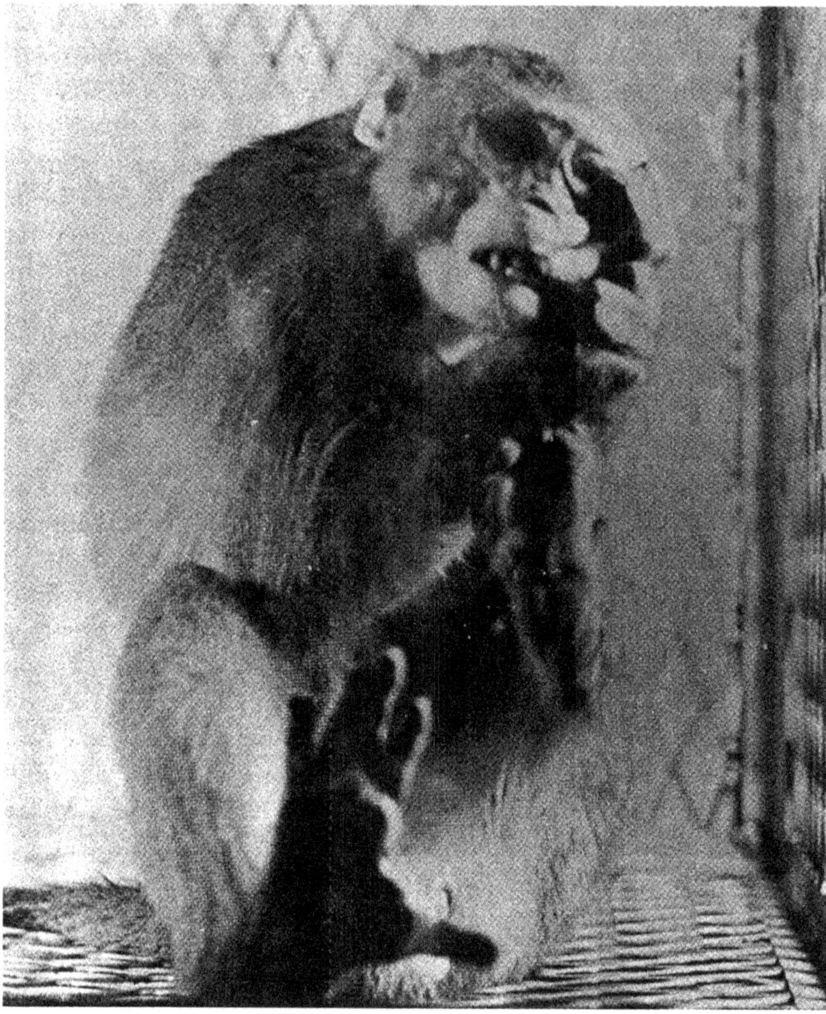

Figure 4. Harry Harlow, *Self-destructive behavior in an isolated male rhesus monkey*, 1974.

This monkey has been weaned in total isolation from all other monkeys and has thus never been mothered, and yet it adopts a posture that is strikingly similar to the pictorial conventions that transmit melancholy affect, one that those studying this monkey have no trouble identifying as an index of emotional distress. Such experiments are no longer conducted today as guidelines for animal welfare and regulations for animal research have altered, partially in response to the very impact of Harlow's findings on the importance of attachment. What detains me here is the clarity and force with which these images capture their viewers and induce instantaneous effects of empathy and concern. Far from demonstrating the primal expressive truth of this posture, such breezy ascriptions of emotional states based upon a basic arrangement of limbs from the scientific observers suggest that it is the very "obviousness" of melancholy posture which requires explanation. If the phenomenon of this obviousness is what we must explain, then the positing of a hidden content, or referential "core," in order to explain that obviousness must be ruled out a priori, for such an explanatory strategy will by its very nature violate and misread that surface. Can we discuss the legibility of the surface without positing a deeper meaning, an underlying content, a propositional assertion that is the truth of that surface?

Thinking of melancholy as posture, and posture as assemblage, offers a relational answer: Melancholy emerges as a structural articulation, an idea which does not transmit a proposition but instead abides as a multiplicity of assemblages, complexly networked into circuits of art-historical dissemination; pictorial convention; and psychoanalytic, psychological, and developmental logics of care and memory. Forming its parts into a clear structure of self-support and self-relation, the melancholy body is an assemblage. But so, too, is the field of relationships between the melancholy postural body and its surrounding material situation, the two forming a constellation of body-earth-objects-things-signs as the self-propping body contemplates and abides with its emblematic and allegorical "props" of shields, blocks, dogs, trees, and rocks. Widening the angle to include both angels and monkeys, one might also regard the archive of melancholy images of these composition-specific, second-order assemblages as itself a tertiary assemblage-of-assemblages, an art-historical array of cumulatively nested structures, each of whose subcomponents articulates melancholy in a locally specific manner. Taken together, all of these melancholy assemblages stand in some relation of consistency to one another: But what is that relation? Is there any particular definitive essence governing their expression other than the notional essence of melancholy itself as the presumed-but-unverifiable causal material fluid within the body which induces these habits and postures, choices of location and object, and scenographic arrangements?[52] Because that fluid cannot be shown and is not, itself, susceptible to presentation, in our visual encounters with melancholy assemblages we are located on a speculative perimeter, provided with an externalized presentation of a material scene. Facing down Dürer's angel and Oliver's Herbert, we are poised to witness a correlationist link of correspondence between the inner and the outer but prevented from verifying the fit between the bodies and things that we can see and the thoughts, spirits, and motions of the soul that we cannot see. As such, the assemblage of the melancholy body offers us the "enigmatic signifier" par excellence, for it arises in a perspectival and limited situation which inescapably raises the possibilities of doubt, failure, and imposture: Behind the gestural surface of the pose, there might be something there that we cannot detect, or there might be nothing there at all.[53] Accordingly, there is something irreducibly proximate about these assemblages, for all their robust and vibrant materiality, and this can trigger a sense that such images give birth to an intolerable, self-satirizing perfume of comedy as easily as they solicit investment and sympathy. The body adopting the posture of a melancholy sage might be frozen in a self-supporting posture of weary despair, or it might simply be bored or pretentiously faking the weariness of intellectual labor out of sheer stupefaction. But the inevitability of the passage from life to death, signaled in the stark difference and tense proximity between live body and dead earth, means that the epistemological instability of melancholy self-presentation rests upon something else within the image: a necrological foundation of dead materiality whose final lingering stasis shall dissolve any and all temporary postures.[54]

The afterlife of this posture could constitute its own book-length study, and indeed Jean Clair's catalogue for the traveling exhibition *Melancolie: Genie et Folie en Occident* constitutes just such a study and just such an archive.[55] But I wish to end this chapter by casting a concluding glance upon a particularly exemplary, contemporary expression of the interpretive dynamic that commences with Adam and Eve's sorrowful posture and redoubles within Dürer and Oliver's paintings. It is an image from the archive of the twentieth-century avant-garde which is itself now captured within the overdetermined narrative of its creator's self-canceling final action: Bas Jan Ader's "I'm Too Sad to Tell You" (1970–1971) (Figure 5).[56]

Now an archival assemblage in its own right, "I'm Too Sad to Tell You" exists in multiple forms: as a lost black and white silent film, a black

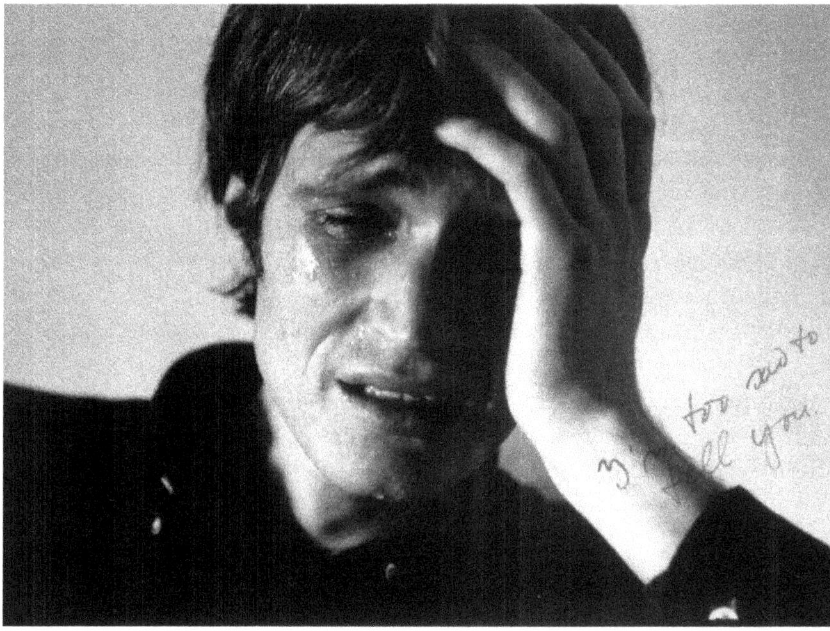

Figure 5. Bas Jan Ader, *I'm Too Sad to Tell You*, 1970. (The Bas Jan Ader Estate, Mary Sue Ader-Andersen and Patrick Painter Editions)

and white photograph in 1970, a series of postcards based upon the photograph and, finally, as a short 16mm film shot in 1971. In each version of the artwork, the representation of sad emotion is perched precariously between contrary readings: The work can be encountered as a direct outpouring of utter sincerity and as a rigged art-historical joke upon that very premise. Whether encountered as a still image or as moving film, "I'm Too Sad to Tell You" short-circuits the very standards of verification and habits of sympathy it also relies upon for its effect. On the one hand, the no-frills presentation of the bodily function of tears offers an insistent, organic authenticity. On the other hand, the tight-framed close-up of both photograph and film evokes the rigged performativity of melodramatic cliché. We cannot know if onions were chopped off-camera or if a painful private emotion was deliberately recalled in order to stimulate the tear ducts to flow on cue; splitting the difference, Jan Verwoert dryly observes that "it is staged but not faked."[57] Regardless of its source, we are brought uncomfortably close to the moment of feeling, yet we have no indication of what, exactly, lies within the face and frame of the artist, racked as he is by seemingly genuine sobs.

The image has been read and reread in relation to the concerns with sublimity and humor that drive all of Ader's work.[58] Here I want to emphasize the legibility and conventionality that the melancholy posture provides to the piece as a historically canny performance of a received idea about how emotion looks. In the second postcard version from 1970, Ader adopts the now-overdetermined, characteristic posture of melancholy and touches his hand to his head as tears stream down his face, effortlessly locating himself within the canonical constellation of the melancholy imaginary already populated by Dürer, Feti, Rosa, and others. In the later, filmed version, crying as process and as action upstages and complicates the melancholic image of grief. Ader's head noticeably rocks back and forth, his mouth becoming an active sensorium, his tongue tasting tears, his lips seeming to hum as he sucks breaths in and out. A transport of emotion sweeps twice across the face, which hovers uncertainly: Is this a grin of mockery or a rictus of total despair? Throughout, the conventional posture of melancholy functions as the "home base," an essential setting; if it centrally defines the still versions of the work, it constitutes an evanescent but repeated tic within the film as Ader covers his face, touches his head, and strokes his hair, repeatedly completing the proprioceptive loop as he adopts and drops the standard pose.

Collectively, all of the existing versions of the piece converge in their shared exploitation of the sly friction performed by the title's mystery upon the embodied facticity of the still/filmed image. Ader's title is a feint, hinting that the "too much"-ness of this sadness has overwhelmed its own transmission: He is too sad to tell us . . . something. How are we to read the declaration that "I'm Too Sad to Tell You" alongside this image, if not as an avant-garde provocation teasing out the confidence trick made available by the basic gap between crying body and thinking mind? In part, we have a splitting of historical difference between an apparently sincere image and a possibly sarcastic title, but the chicken-and-egg question of priority only doubles the problem of intent already implicit in the image itself. Acting upon Ader's painting much as the "Magia Sympathia" slogan acted upon Oliver's painting, the title indexes the essential epistemological gap between interior cause and exterior outcome. Once noticed, the shadow of artifice and essential theatricality raised by the histrionic nature of the "too much-ness" implicit within the title proves very difficult to dispel, and not simply because that phrase is scrawled directly upon the image. Crucially, the title works to impress the viewer into an interpretive duty of care, concern, and stance-taking toward the image from which, at the very same time, they have also been categorically excluded. We are given to know that we do not know what the melancholy figure knows about their own emotional state. Affect and epistemology are thus braided together and resolutely separated at the same time: The undeniable material reality of tears seemingly proves the very sincerity of feeling as a bodily event while the title belies and disavows any secure form of knowledge about the tear's occasional cause which the viewer could verify for herself. Expanding the frame to consider the receptive circuit in which artwork and viewer are brought into relation through the aggressively preemptive mediation of the title, one might parse the resulting dynamic as a formula: "I" + the title + viewer as "you" = the "work" as a total relay of conjunctive disjunction. Together these components form a larger melancholy assemblage of affective flows and interpretive energies that feed forward and back across the very boundaries of access mobilized by the title's prickly declaration.

There are locally situated narratives here about the state of Ader's career, his personal history, his relationship to West Coast art-making strategies, and the contexts in which his work circulated then and still circulates today; but against historicist capture, I would maintain that the real virulence of this particular self-consuming image already lies dormant within earlier examples from the longer arc of melancholy representation, where the unverifiability of the emotional reality attested to by melancholy posture emerges from the frame upon each viewing, similarly addressing and refusing the viewer's participatory engagement and emotional solidarity. If the mystery attendant upon the ur-source of Dürer's angel and the difference between Herbert-as-sitter and Herbert-as-man offer any indication, self-differential disjunction is always already present within the art-historical archive of melancholy portraiture. Taken together as a trio, Dürer's angel, Oliver's Herbert, and Ader's canny self-portrait offer us a polychronic melancholy assemblage in which the ambivalently postured melancholy body functions as a persistently incorrigible solicitation, one which swirls together equal parts sympathy and suspicion. The longer one looks at Ader's work, the harder it is to determine whether we are being encouraged to weep with or laugh at its charged spectacle. It is this unresolvable tension between the comic and the terrifying which also drives Shakespeare's priming encounter with melancholy assemblages as dramatic situations in *Love's Labour's Lost*, to which I shall now turn.

2. *Three Hundred Years Out of Fashion*

Affect and Affectation

Love's Labour's Lost has irritated playgoers for centuries. Coming to grips with this slippery, curious play in his *Remarks on the Plays of Shakespear* (1710), Charles Gildon first pulls, then throws, his punch: "[. . .] [S]ince it is one of the worst of Shakespear's Plays, nay I think I may say the very worst, I cannot but think that it is his first, notwithstanding those Arguments, or that Opinion, that has been brought to the contrary."[1] One hundred years later, Hazlitt demurs on the date but concurs in his review: "If we were to part with any of the author's comedies, it should be this."[2] What makes Shakespeare's fashionably witty first foray into the dramatic representation of melancholy so singularly off-putting?

The play begins with a king gripped by the fear of death and ends with the news that a king has died; what falls between is comedy, but comedy of a curiously static kind. Courtiers and fashionable fools chat and dance, some letters are sent back and forth, a masquerade flops and ends with a hasty farewell, but there is very little narrative suspense or properly dramatic action. Replete with verbal pyrotechnics and aureate wordplay, *Love's Labour's Lost* is a comedy in which a great deal is said and yet, in comparison with the transformations of status and unforeseen punishments and rewards typically brought about by Shakespeare's comedies, in this play one is tempted to say that "nothing happens." This retreat from action into language places enormous pressure upon that language: It must take on the status of action. It is a shift that follows directly from the King of Navarre's decree that opens the play:

KING: Let fame, that all hunt after in their lives,
Live registered upon our brazen tombs,
And then grace us in the disgrace of death
When, spite of cormorant devouring time,
Th' endeavour of this present breath may buy
That honour which shall bate his scythe's keen edge
And make us heirs of all eternity.
Therefore, brave conquerors—for so you are
That war against your own affections
And the huge army of the world's desires—
Our late edict shall strongly stand in force.
Navarre shall be the wonder of the world.
Our court shall be a little academe,
Still and contemplative in living art. (1.1.1–14)

The grandeur of the King of Navarre's opening lines prepares us for a stoic repudiation of worldly fame, which makes it all the more comic that this speech is in fact a proposal for how to secure it. Far from abandoning the world and its desires, in promising that "Navarre shall be the wonder of the world," the King comes across as transparently eager. But fame is not his ultimate goal. Rather, fame functions here as a means to stay the "scythe's keen edge." Yet the remedy that Navarre advocates (keeping "still and contemplative in living art") seems dangerously similar to the very threat of death and effacement that the glory of the "little academe" is meant to ward off. If the unraveling of Navarre's academy of heroic masculine renunciation in the face of feminine temptation constitutes the play's comic payoff, there remains something awkwardly static about the experience of a play frozen in verbal contrivances. To borrow a contemporary diagnostic phrase, there is something manic-depressive about *Love's Labour's Lost*. The glittering, manic surface of volubility and courtly manners that makes up the ostensible action of the play works to ward off the depressive pull toward "still and contemplative" paralysis at its core. The play constitutes a series of manic entertainments that struggle to defer the final, inevitable plunge into depression.

Offering both a fashionable courtly identity and an explanatory frame for morbid anxiety, melancholy haunts both halves of *Love's Labour's Lost*'s manic-depressive structure and proves malleable enough to figure prominently in both its distracting entertainments and in its final disclosure of the fact of death. To do so, Shakespeare draws upon opposing elements from melancholy's divided intellectual history. The men of Navarre's court, in particular Armado and Biron, fashion themselves as melancholic geniuses after the Aristotelian model of "genial melancholy." By contrast, the account of Katherine's sister's death from "heaviness of heart" radically refigures melancholy not as a discursive affectation but as a deadly serious kind of affect, a depressive illness more consistent with the teachings of Galenic medicine. Poised between these conflicting medical models, melancholy is at once a merely verbal expression of foppish male pretension within a broader satiric gallery and a somatic touchstone of female steadfastness and authenticity that provides critical leverage against the impostures of Navarre's "little academe." The persistence of this discursive overdetermination within Shakespeare's play could be taken as typically "Renaissance" syncretism or, less flatteringly, as a failure to choose, but it performs a crucially unifying function in manifesting the construction of melancholy as a gendered and gendering assemblage. Restlessly shifting from the comically "light" to the tragically "heavy" and back again, in *Love's Labour's Lost* melancholy functions as a kind of switch point along the abyssal divides between affect and affectation, men and women, minds and bodies, and public expression and private interiority, suturing together the play's manic entertainments and its depressive closure. What begins with the convention-bound recognition of the melancholic as a fashionable social type, ends as a skeptical encounter with melancholy as an unknowably encrypted internal space; but that aporetic encounter both shatters and reforms the play's affective networks within and across boundaries of gender, national identity, and class. As a stance fraught with anxiety about its own impersonation, melancholy may not (quite) verifiably exist, but its provisional status grants it a curiously effective form of "weak" power to reconstitute partnerships, friendships, marriages, and obligations of servitude and obedience nonetheless.

Don Adriano de Armado, Walter Benjamin, and the Dialectical Theater of Melancholy Fashion

A foil even to other foils, Armado's peculiar status puts him continually at odds with all of his companions, both within the court and outside it. As such, he offers a compelling model for the way that melancholy simultaneously dissolves and solidifies social networks of affinity and belonging, allowing "inside" and "outside" to touch. Distinguished from all other characters by his nationality, his speech patterns, and his penchant for personal display, Don Adriano de Armado functions as a kind of emissary for difference itself. As a Spaniard in Navarre, his difference carries a nationalist charge and his name a topical reference to the Armada, and this note sounds whenever his name surfaces in courtly conversation. The King's first introduction of him ("Our court, you know, is haunted / With a refinéd traveller of Spain, / A man in all the world's new fashion planted" (1.1.161–163)) manages to render Armado as both a "man of the world" and, specifically, Spanish. Yet his status as a gentleman, if one of limited means, puts him at odds with the other so-called grotesques with which he is customarily grouped, as evidenced by his employment of Moth, arrest of Costard, and self-consciously condescending love for Jaquenetta (whom he casts as beggar to his King Cophetua). Executing every customary gesture of courtly politesse with a flourish of manic gusto, Don Adriano de Armado's attempts to charm and impress the academic court of Navarre go perfectly wrong, with richly comic results. His "eloquence" and "learning" may amount to comic bombast and torturously mangled phrases, but the very excess of matter and paucity of meaning in Armado's speeches and letters provide exemplary entertainment for the scholarly King, his circle of gentlemen acolytes and, by extension, the audience. The King's initial description of him as a "child of fancy," and "one who the music of his own vain tongue / doth ravish like enchanting harmony" (1.1.170; 165–169) prepares us to receive Armado's *amour propre* as charming rather than repellent. Warming up the crowd, so to speak, the King's opening description of Armado's narcissism reads like a grotesque parody of the capacity for self-overhearing that Harold Bloom has polemically defined as Shakespeare's signature literary achievement.[3] Armado can *only* hear himself, and so his self-directed "minstrelsy" generates a curious extension of Bloom's dynamic, as his own narcissism turns those around him into over-hearers as well.

Armado's reputation for inflated linguistic performance precedes him; he enters the play as a comic toy of discourse before showing up in person, and from the start it is his earnest, grandiloquent declarations of melancholy that set him up as a fool. Armado's penchant for proclamations of his melancholy and his penchant for pretentious linguistic excess are introduced simultaneously when Ferdinand reads aloud from Armado's letter detailing the causes of Costard's arrest at the end of the play's first scene: "So it is, besieged with sable-coloured melancholy, I did commend the black oppressing humour to the most wholesome physic of thy health-giving air" (1.1.227–228). Armado says what goes without saying. Melancholy is, by definition, black in colour, but Armado, inflating and extending language wherever possible, clads melancholy in the sufficiently regal and aureate epithet "sable-coloured" and then subjects it to a further gloss as a "black oppressing humour." Armado's language betrays his doubled status as both courtly and "grotesque," and places him squarely between the play's opposed worlds. If his relentless linguistic invention and turnings of phrase aspire toward the lightning strikes of courtly wit typical of Biron, Armado's love of tautological verbal padding lends his utterance that curious sense of traveling without moving that is the hallmark of the pedant Holofernes. If nothing is added to melancholy in glossing it as black oppressing humour at the level of reference because the two terms designate the same thing, nonetheless something *is* added at the level of connotation. Armado's proliferation of equivalent epithets plays with identity as a formal principle, even as it sketches his own personal identity through a repetitive entrapment in a description of his inward suffering. While parsing such trifles at this level risks participating in a similar act of linguistic inflation, the very paucity of content and formal archness of Armado's discourse are both essential to the production of the melancholic repertoire of bewitchingly blank surfaces and stagey, blurred withdrawals from content. Armado occupies a position of empty fullness, ceaselessly additive but without a final result. The assertion of melancholy, a condition which one expects might shut down language, instead functions here as a cue for further, potentially endless, elaboration. This dynamic of proliferation out of poverty—and poverty within proliferation—defines the melancholic threat lurking beneath the "great feast of language" which makes up *Love's Labour's Lost.* Costard's lament about the vapid exchanges of Holofernes and Dull that "they have been at a great feast of language and stolen the scraps" (5.1.35) also articulates the paradoxical empty fullness of Armado's melancholy bathos.

Warding off this threat of emptiness by presenting himself as a "man of greatness," Armado's claim to suffer from melancholy should be read as a strategy for securing status. The constellation of melancholic suffering with exceptional talent or ability is a commonplace that originates, as we have seen, in the pseudo-Aristotelian textual tradition of "genial melancholy" and the pretentions and imitations it induced. Casting its nets widely across the varieties of human achievement but remaining tantalizingly evasive on the causal directionality between the twin poles of melancholy and genius which it binds together, the very vagueness and breadth of "Problemata XXX.1"—combined with the powerful gravity exerted by its supposed source in the preeminent textual authority of ancient times—proved astonishingly generative. In a sense, Don Armado is both a product of the pseudo-Aristotelian question's hold upon the early modern cultural imagination and a parodic proof of its deeply familiar, shopworn status by the late sixteenth century. While one can certainly doubt the extent to which a single textual passage could be responsible for an entire intellectual tradition spanning centuries and national and linguistic barriers, there can be little doubt that the association of melancholy with genius was a suitably familiar received idea by Shakespeare's time. The spectacle of someone hoping to achieve genius by falsifying melancholy could only be humorous to an audience already presumably familiar with the doctrine of their supposed overlap.

In his first appearance in person upon the stage, Armado tensely reasserts his letter's self-diagnosis and tries to displace the question of whether he is actually melancholy at all by posing the leading and potentially unanswerable question "*why* am I melancholy?" In pseudo-Aristotelian fashion, this assertion of melancholy presumes a prior evaluation of personal "greatness":

> *ARMADO:* Boy, what sign is it when a man of great spirit grows melancholy?
> *MOTH:* A great sign, sir, that he will look sad.
> *ARMADO:* Why sadness is one and the selfsame thing, dear imp.
> *MOTH:* No, no, O Lord, sir, no.
> *ARMADO:* How canst thou part sadness and melancholy, my tender juvenal? (1.2.1–8)

Shakespeare will return to this scenario (a man asking another man to interpret the cause of his sadness) with greater subtlety and dramatic power at the beginning of *The Merchant of Venice*; here it functions primarily to provide satirical leverage against male narcissism. Armado is soliciting an interpretation of his emotional interior, and the comedy of Moth's reply lies in its insistence upon surfaces rather than depths. For Moth, melancholy is a matter of looking sad rather than being sad, and so he transposes "greatness" from a quality of men to a quality of signs. In a sense, this is simply consistent with the other subplot characters who continually subject Armado's rhetoric to comically literalist, deflationary readings, as in Costard's translation: " 'Remuneration'! O, that's the Latin word for three farthings" (3.1.132).[4] In cashing out melancholy as a matter of appearances rather than emotions, Moth is not contradicting Armado's assertion that he is a melancholic but redefining what that state consists of. As G. R. Hibbard notes, Armado's opening line is a purely rhetorical question, fishing for the answer "it is a sign that you are in love," in keeping with an Ovidian commonplace on melancholy as a symptom of love: "Ovid, in his *Ars Amatoria*, i. 737–8 writes *'Ut voto potiare tuo, miserabilis esto; / Ut qui te videat, dicere possit, Amas.'* "[5] Don Armado's question "how canst thou part sadness and melancholy" is itself left unanswered as Moth and Armado fall to semantic quibbling over the epithets they use to designate each other. The resulting conversational drift away from Armado's supposed depression into displays of wit, however pretentious or feeble, suggests that he fails to properly exemplify even Moth's reductive definition of the melancholic, for there is nothing particularly sad-seeming about Armado's wordplay. In the process, this exchange troubles the distinction between the play's primary characters and these denizens of the subplot, for the rapid fire exchanges between Armado and Moth echo the banter that takes up the Navarre court's mental energies.

To detach Armado's question from its local context, how *can* sadness and melancholy be parted? This parting of sadness from melancholy is, in a sense, what the narrative energy of the play itself will eventually achieve when the death of the King of France forces the separation of the bereaved court of the Princess from the love-melancholic court of Navarre. At that point, Armado will answer his own question, for it is his voice which closes the play by placing a seal upon precisely this parting with a decisive "You that way. We this way" (5.2.912).

Shakespeare is both demonstrating his knowledge of melancholy's dramatic conventions and, in foregrounding its very familiarity as a received idea, critiquing the basic assumptions upon which the legible social type of the melancholy genius relies. Of course, the fact that some people pretend to be melancholic in order to seem to be great does not logically prove that "truly" great people are not "truly" melancholic. But Moth's redefinition of melancholy registers impatience with melancholy itself, and not just with supposedly false performances of it. Playing Armado's posture of melancholic greatness for laughs, Shakespeare uses a dramatic convention as a tool for folding social knowledge back upon itself as he moves the audience from facile recognition to doubt at whether there was ever anything there to recognize in the first place. If it seems counterintuitive to suggest that pseudo-melancholics might be recognizable *without reference to real melancholics*, such a vanishing of the supposedly original standard of judgment suits an intellectual tradition founded upon a textual experience of misrecognition in which the writings of Theophrastus are taken to be the work of Aristotle.

Such a transformation exhibits one of *Love's Labour's Lost*'s recurring strategies, namely the harrowing of knowledge, in which postures of intellectual mastery are shown to enfold stupidity as a kind of necessary lining or encrypted truth. This is manifested structurally in the correspondences and pressures of resemblance between the members of court and the "grotesques" who make up the play's subplot.[6] Seen in this way, Holofernes' pedant functions as the grotesque realization of the King of Navarre's own scholarly ambitions. Exemplifying the pursuit of scholarly knowledge for its own sake, Holofernes' conversations register as the sputtering of a perpetual discursive motion machine cut off from any possible application or relevance, both manically communicative and yet ultimately locked in narcissistic self-regard.

Armado's melancholic minstrelsy is also a posture of knowledge resting upon a foundation of enabling ignorance, and while much of this constitutive blind spot can be chalked up to ingrown affectation, discursive "bad habits," and a xenophobic caricature of national identity, there is also a sense in which Armado's knowing ignorance is a matter of class. If Moth's responses relocate the identity of the masculine love melancholic from its unseeable depths to its socially legible surfaces, it is hardly accidental that he does so as a servant addressing a master.[7] Drawing on a comic tradition at least as old as Plautus, such repartee is not necessarily subversive, but it does further reinforce the association of love melancholy with a particular kind of class membership, and the distance that membership entails is rhetorically capitalized on by his wisecracking servant. Using his cultivation of all that is "fire-new" and in fashion in courtly circles against him, Moth brilliantly exploits the extent to which Armado's immersion and familiarity with one kind of social knowledge excludes him from other discourses, and in particular the kinds which might actually lead to a successful wooing of Jaquenetta:

> *MOTH:* Master, will you win your love with a French brawl?
> *ARMADO:* How meanest thou? Brawling in French?
> *MOTH:* No, my complete master; but to jig off a tune at the tongue's end, canary to it with your feet, humour it with turning up your eyelids, sigh a note and sing a note, sometime through the throat as if you swallowed love with singing love, sometime through the nose as if you snuffed up love by smelling love, with your hat penthouse-like o'er the shop of your eyes, with your arms crossed on your thin-belly doublet like a rabbit on a spit, or your hands in your pocket like a man after the old painting; and keep not too long in one tune, but a snip and away. These are compliments, these are humours, these betray nice wenches that would be betrayed without these; and make them men of note—do you note me?—that most are affected to these. (3.1.9–23)

Moth's mocking query about the French brawl taunts Armado with its own obscurity. In a mixture of rural and urban registers, Moth is in fact describing a country dance in language charged with images borrowed from city fashions, and in particular with images derived from the symptoms of fashionable love melancholy. The dance in question has been ably glossed by A. S. Moncure-Sime in *Shakespeare: His Music and Song*:

> The brawl was one of several tunes to which the Country Dance was danced, whether in a ring, or 'at length', like our 'Sir Toby'. Brawl was the English of the French 'bransle' or 'branle.' Like the Allemande of Bach, 'it containeth the time of eight, and most commonly in short notes'.[8]

In the process of explaining "the French brawl," Moth sketches a rapid-fire anatomy of love melancholy, detailing its effects on the tongue, feet, eyelids, throat, nose, eyes, arms, belly, and hands. Moth's anatomy consists not of somatic symptoms caused by a substantial illness but of citational behaviors expressive of a social type. Simultaneously jigging, sighing, singing, sniffing, and smelling, this lover joins hyperkinetic activity with performative inadequacy ("not too long in one tune, but a snip and away"). While certain details of this passage suggest a correspondence with contemporary visual images of the melancholy lover (e.g., the depiction of "Inamorato" on the frontispiece of Burton's *Anatomy* or the portrait of Donne as a melancholic lover complete with floppy brimmed hat), Shakespeare here exploits language's capacity for pictorialism while exceeding the consistency of a single visual image. In the tumbling addition of mutually contradictory details, Moth fashions an impossible chimera: Why would one turn up one's eyelids to heaven while *also* wearing a hat brim low over one's face? Through his exploded view of the gestures and tics that make up the correct performance of the country dance, Moth is here rehearsing the conventions of love-melancholy as an essentially manic activity haunted by the depressive fear of failure. He does so in a manner calculated to induce both anxiety and imitation in Armado. Moth's litany of amorous melancholic modes both reflects knowledge of the love-melancholy convention and, in turn, amplifies and recirculates that convention through this self-same public assertion of its familiarity—a familiarity which marks the conventions of love melancholy as somehow both dated *and* fashionable.

In his influential preface of 1927, which commenced the twentieth-century revival of dramatic and critical interest in *Love's Labour's Lost*, Harley Granville-Barker wrote: "Here is a fashionable play; now, by three hundred years, out of fashion."[9] Yet Moth's very mockery of the conventions of love-melancholy within the play suggests that some of the matter within might not have been so very fashionable after all. Reading Granville-Barker's preface together with Moth's disdain lodges the play within a chiasmus of redoubled temporality that disarticulates the logic of fashion: The unfashionably fashionable play of 1595 is resurrected before our eyes as the now-fashionably unfashionable play of 1927. In a sense, Granville-Barker's preface only reinscribes Moth's own manipulation of what, granting apologies to Foucault, I would like to call fashion-knowledge. Both speakers invoke the spectre of the unfashionable under the pretense of deconstructing and mocking the dynamic of fashion by stressing its transience, while in effect reinforcing and redeploying that very dynamic for their own rhetorical ends. The temporal paradox of fashion-knowledge lies in the citational requirements it places on recognizability: To be fashionable is to be of the moment, not so far ahead of it as to be unrecognizable as fashionable, nor so far behind as to have fallen out of popularly perceived taste. But in order to perceive something as fashionable, we have to perceive it as a citation of something that has been seen before or recognition of its very fashionability could not take place. This requirement shifts the fashionable status of any given object ruthlessly toward its term limit at the very moment its fashionability becomes apparent. This dynamic suspends certain poses, behaviors, garments, manners, and attitudes in a kind of hovering morbidity of quasi-obsolescence even as it confers upon them the status of being "au courant." In a remark from the *Passagen-Werk*, Benjamin describes this interpenetration of past and future within fashion as a dialectical process:

> Each time, what sets the tone is without doubt the newest, but only where it emerges in the medium of the oldest, the longest past, the most ingrained. This spectacle, the unique self-construction of the newest in the medium of what has been, makes for the true dialectical theater of fashion.[10]

Moth's manipulative use of his own fashion-knowledge against his master demonstrates the flux and mutability at work in such a dialectical theater: City dandies jig country dances and today's fashions and yesterday's conventions are opposed only to be shuffled into a coextension that transvalues both.[11] Moth's exchange with Armado thus allows us to track the multiple timelines along which the social type of the male love melancholic circulates: The gestures and modes grouped under the "French brawl" are passing out of fashion and into convention for Moth precisely at the moment that he transmits and commences their fashionable, and thus ostensibly preconventional, status for Armado.

Its presentness always haunted by the incorporation of the past, fashion itself would seem to be the site of a particularly melancholic haunting, since it consists of techniques for the cultivation of the self which can be achieved only through the citation and manipulation of past fragments whose very evocative power continually threatens to overthrow the hold of fashion's eternal "now." Metadiscourse about fashion inherits the same temporal distortions and hauntings. In his foreword to *The Fashion System*, Roland Barthes, remarking upon the explosion of interest in and development of "semiology" between 1957 (when he began his project) and 1963 (when it was published), laments that: "This venture, it must be admitted, is already dated."[12] It is as if there is something slightly toxic about contact with the temporality of fashion which forces the acute risk of embarrassment onto those who comment upon it. If it is true that "semiology" had changed between 1957 and 1963, so had the climate of fashion that Barthes was describing; in the wake of the "New Look" of Courréges circa 1963, the entirety of Barthes' exemplary material would have seemed rather dowdy. Barthes' shame at the "datedness" of his own working methodology should be read as a symptomatic repetition of the temporal drama of fashion which makes up his (base) subject matter within *The Fashion System* itself, confirming Benjamin's insight into the dialectical nature of fashion-knowledge. When grasped as a bitter temporal struggle for self-definition rather than as simply a game of quotation and influence, this dialectic of fashion can be seen to exhibit the same dynamic which determines our experience of *Love's Labour's Lost*: manic turnover and ceaseless novelty that wards off (yet maintains) a depressive, malingering paralysis.

Biron, Sigmund Freud, and Melancholy Wit

The overproductive surplus of personal display associated with melancholy self-fashioning (to use, suitably, a once fashionable, now passé critical term) has long been noted. In "Mourning and Melancholia," Freud notes with ironic surprise that:

> It must strike us that after all the melancholic does not behave in quite the same way as a person who is crushed by remorse and self reproach in a normal fashion . . . One might emphasize the presence in him of an almost opposite trait of insistent communcativeness which finds satisfaction in self-exposure.[13]

If this insistence seems definitive of Armado and his minstrelsy, it is equally true of Biron's "sweet and voluble discourse," and yet the play seems to allow their ultimate resemblance to take shape only gradually.[14] From his first appearance, Biron seems defined as the true wit in opposition to Armado's false wit. Biron's rhetorical set pieces extolling temperance and harmony, his "like of each thing that in season grows" (1.1.107) would seem to define him as the opposite of Armado's intemperate excesses of passion and pretense. Indeed, Biron's sarcastic praise of Armado's letter as "the best that ever I heard" and his mock encomium of the Spanish gentleman as "a most illustrious wight / A man of fire-new words, fashion's own knight" (1.1.268, 76–77) betray no little contempt. Yet those very words, if delivered "straight," could just as well designate Biron as they here designate Armado when delivered with Biron's cuttingly polite manner. As the play progresses, these two characters are brought into circulation around each other, their resemblance intensifying as their passions and melancholy take hold, and finally come to occupy each other's positions with the mistaken switching and recirculation of Armado's letter for Jaquenetta and Biron's letter to Rosaline. If Armado is the narcissist ravished by his own music, Biron uncannily doubles him, for his own attempts at lovemaking lead ceaselessly back toward himself, and his own wit:

> BIRON: Learning is but an adjunct to ourself,
> And where we are our learning likewise is:
> then when ourselves we see in ladies' eyes,
> Do we not likewise see our learning there? (4.3.311)

These displays of pseudo-syllogistic sleight-of-hand are meant to display Biron's skill at argument, but they tend to hinge upon sophistic equations that undercut their ultimate persuasive force. Like the reflected vision of himself that he encounters within his mistress' eyes, Biron's learning lies upon the surface. These exertions of rhetorical athleticism are clearly successful as social calling cards, for Rosaline glowingly recalls Biron's "sweet and voluble discourse," but they are somewhat less successful as arguments, as he learns in his actual courtship of her.

Courtship is, among other things, an act of rhetorical persuasion and requires that one prove one's love to be genuine. For both Biron and Armado, melancholy here functions as evidence to support their love suits; but since melancholy itself requires proof, its grounding force is always in doubt. In the world of *Love's Labour's Lost*, there is a mutual relationship of evidentiary support between the claiming of both states ("I am in love, and therefore it has made me melancholy"; "I am melancholy, and this must mean that I am in love"). Neither can quite prove the other, for each refers back to the other for support. Conventional belief in love as one of the primary causes for melancholy only completes the fatal circularity of an infinite regress. Comedic courtship need not be logically ironclad; what is significant within the link between being in love and being melancholy is the experience of their discursive mingling and disturbing resemblance.

Both Biron and Armado experience melancholy not as an illness in its own right but primarily as a symptom of a deeper, originary state—that of being in love. Echoing Armado's Ovidian belief that love causes melancholy, Biron regards his melancholy as a symptom of his being in love, for love has "taught [him] to be melancholy"; such an etiology was simply a given of the medical and psychological thinking of the period, but note that Biron's pedagogical metaphor reinforces a discursive rather than medical understanding of the illness. The effect of love melancholy on both Armado and Biron is also similar, and just as discursively oriented: both write complaints addressed to the supposed causes of their state which detail their author's painful fidelity and afford generous opportunities for the display of courtly civility and wit. Anticipating the country gulls and city gulls of Jonson to come, Armado brays that "I am sure I shall turn sonnet. Devise, wit; write, pen; for I am for whole volumes in folio." (1.2.174); writing falls immediately after loving on Biron's "to do" list for erotomaniacs: "Well, I will love, write, sigh, pray, sue and groan" (3.1.197). What is fundamentally similar here is the glibly automatic self-control with which they motivate themselves to further limn their own supposed suffering. Far from registering an incapacitating passion, love melancholy is above all here figured as active, textually generative, an incitement to further action rather than the sluggish retreat into self which one might expect. In this sense, the play bears out Moth's critique; by demonstrating the haste with which Biron and Armado seek to demonstrate their sadness for others, melancholy functions as a "great sign of seeming sad," exerting an imperative pull toward communication rather than a retreat into the self. As the index of a painful gap between lover and beloved, love melancholy would seem to denote isolation and separateness. But as an expressive act of "love speech," whether directed toward its object-cause or indirectly voiced as a means of securing some leverage of sympathy from onlookers and bystanders, the presentation of love melancholy binds persons together and anneals the very wound it also announces.[15]

Biron's love for Rosaline constitutes the violation of a sworn promise, and a promise made to his sovereign. If *Love's Labour's Lost* really were a tragedy, then Biron's violation would lead to a stark punishment, as tends to happen in folktales which hinge upon "the rash vow," which is the essential narrative structure underlying the framing device of the King of Navarre's scholarly retreat. But Shakespeare's comedy delays the consequences of this violation, using the spectacle of Biron-in-love as the occasion for a testing of the limits of conventional love language by placing a character determined to display a discourse not only "sweet and voluble" but *original* into the most cliché prone state imaginable. Biron's self-conscious disgust at being in love reveals his alienating awareness of its conventionality, of the given-ness and artificiality of its language and rituals. This situation is particularly painful for a man of wit. Precisely when he is most moved to articulate an emotion, he finds that it has been so ceaselessly represented and enacted by others before him that the possibility of generating anything new to say on the subject simply disappears. Love is soiled by discursive circulation. Caught between lyrical expression and a corrosive skepticism, in his soliloquy Biron plays both Moth and Armado:

BIRON: And I, forsooth, in love! I, that have been love's whip,
A very beadle to a humorous sigh,
A critic, nay, a night-watch constable,
A domineering pedant o'er the boy,
Than whom no mortal more magnificent!
This wimpled, whining, purblind, wayward boy,
This Signior Junior, giant dwarf, Dan Cupid,
Regent of love-rhymes, lord of folded arms,
Th' anointed sovereign of sighs and groans,
Liege of all loiterers and malcontents,
Dread prince of plackets, king of codpieces,
Sole imperator and great general
Of trotting paritors—O my little heart!—
And I to be a corporal of his field,
And wear his colours like a tumbler's hoop! (3.1.167–181)

Like the hyperkinetic melancholic who dances the French brawl in Moth's teasing excursus to Armado, Biron here fashions not so much a recognizable portrait of a social type as a manically amplified catalogue of every possible expression of masculine infatuation. In the midst of this torrent of epithets, the very simplicity of Biron's exclamation "O my little heart" erupts. If this cry belongs with the "russet yeas and kersey nos" which he will later commit to for Rosaline's sake rather than the "taffeta phrases" he is here busy cutting and sewing into place, it is no accident that it should be a kind of bodily cry of pain which arrests the ceaseless tumble of dependent clauses of abuse. And yet there is no independent way of experiencing and verifying that this cry of "O my little heart" has any claim not to be included within the scope of Biron's contempt for "whining" and the "sighs and groans" of others. This may be a case of dramatic irony achieved through the humbling of a solipsistic snob who regards the loves of others as mere convention while insisting upon the integrity of his own passion by crying out uncontrollably. But it might also occasion a more radically skeptical assertion on Shakespeare's part of the necessary limits upon representation faced even by someone as agile with language as Biron. The text insists upon bodily suffering as the limit of language.

This intrusion of a signal toward an interior of suffering made within language but seeming to gesture beyond it reoccurs throughout Biron's soliloquies and asides. Biron's contempt for love drives him to resolutions against it which only occasion further concessions to its power and confessions of failure, and repeatedly that moment of passive surrender announces itself in an inward turn toward the heart or breast.

> BIRON: I will not love. If I do, hang me. I'faith, I will not. O, but her eye! By this light, but for her eye I would not love her. Yes, for her two eyes. Well, I do nothing in the world but lie, and lie in my throat. By heaven, I do love, and it hath taught me to rhyme and to be melancholy; [showing his paper] and here is part of my rhyme, [pressing his hand to his breast] and here my melancholy. (4.3.7–14)

Biron's speech articulates a tense dynamic between the promissory language of rational will and the destructive effects of desire, which he figures as an ineluctable force. The speech commences with the declaration "I will not love," which means here "I am not going to feel love," but could also mean "love is not something I can control by my will." This already-shaky declaration is advanced only to be qualified immediately by the possibility that it will be violated in Biron's "If I do," and so the penalty of hanging is advanced to place a mortal seal upon his resolve. While the expression is proverbial in a play saturated with proverbs, when Biron briefly imagines death by hanging as a punishment for love he sounds a theme that will be repeated and transformed as the play proceeds. Later, in a locally specific reference to the Tyburn gallows, Biron laments "Love's Tyburn that hangs up simplicity" (4.3.52); in the final act, Katherine also describes Cupid as a "shrewd unhappy gallows" (5.2.12). Biron's willingness to suffer a capital punishment for the breaking of a promise is another example of the conventional language of oath-taking which still survives in the schoolyard oath "cross my heart and hope to die." But this additional security is offered only to be immediately cancelled by the recollection of Rosaline's eyes, comically undermining the language of declaration with a passionate lament. The shift of the modal verb tense from "I *will* not love" to the cry "but for her eye, I *would* not love her" (italics mine) figures Biron's love for Rosaline as an undesirable, but inescapable, fait accompli.

In pointing to his paper and then to his breast, Biron proffers a kind of forensic inventory of objective evidence for the claim "I do love." There is a deictic specificity to Biron's gestures which is meant to reinforce their persuasive force: My rhyme exists, here on *this* paper, and my body exists, it is *this* body here before you.

> BIRON: By heaven, I do love, and it hath taught me to rhyme and to be melancholy; [showing his paper] and here is part of my rhyme, [pressing his hand to his breast] and here my melancholy.

By shifting from oaths made with proverbial, received expressions ("in faith," "by this light," "by heaven") to an ostensive gesture toward *this* paper and *this* body, Biron tries to drive some kind of barrier between the instability and potential falsehood of promissory language and the ontological bedrock of thingly existence, and to pledge the faithfulness of his love on the self-evident thingliness of the paper and the undeniable presence of his own body. Eliding the discursive and the somatic, Biron's survey of evidence alleges a similarity between the paper and the breast as equally demonstrable, and therefore equally persuasive, proofs of love. This elision is of rhetorical convenience, for it manages to make the breast metonymically stand in for the melancholy allegedly contained within it, and the ontological security with which the breast can be thumped upon and pointed at is meant to secure an epistemological security about the emotions which supposedly reside there. The problem is that this transmission can only happen if the audience already trusts and believes that Biron is melancholy, and so the gesture cannot establish that belief itself. If I say, "I have the money right here" and point to, but do not open, my wallet, your belief that I do or do not have the money will be a register of your prior trust in me, and nothing more.

This bears further explanation. The idea that the breast specifically should function here as a touchstone of truth is particularly important in a context in which Biron seeks to distinguish his feelings of genuine love melancholy from the mere pretensions of "fancy" (which would have been understood to be located in the brain). The precise location of melancholy as an illness had long been subject to debate by medical authorities, but there was a popular belief that feelings of romantic passion, and the sadness attendant upon its frustration, were localized in the heart.[16] French physician Jacques Ferrand conveys the hold of this popular consensus when, in his *De la Maladie d'Amour ou Melancholie Erotique* (1610, translated into English 1623), he begins his chapter entitled "Whether in erotic melancholy the heart or the brain is the seat of the disease," with the straightforward testimonial: "If you question our lovers concerning the part wherein they feel the greatest affliction, they will all tell you that it is the heart."[17] While the author goes on to compare claims for the authority of both organs from the pantheon of contemporary and ancient medical sources, with the heart finding its champions in Avicenna and pseudo-Hippocrates, what is more important for our purposes is the popular sense in which the heart is clearly perceived as the natural location for feelings of love and lovesickness. Language, publicly sealed and recognized, calls for some kind of touchstone of integrity and validity, and here it is signified by the physical interiority of the breast and its popular definition as the privileged location for the passions Biron claims to experience. However, the thoroughly conventional nature of the gesture introduces a generality that subverts its deictic specificity and its persuasive force. Biron may be swearing by his breast, but that gesture is itself a legible sign that might or might not be feigned along with all the other "sighs and groans" and "folded arms" that he mocks in others. To raise this doubt is perhaps to raise Hamlet's hermeneutics of skepticism with regard to the "actions that a man might play"; here they remain ghostly traces of doubt, implicit but as yet unspoken.

Such concern for the force of precedent rearticulates the temporal struggles between old and new which drive the "dialectical theater of fashion" in a more anxiously personal register; in conveying their love complaint, melancholics seem caught between the given-ness of the received form and its urgently specific expression in the "now" of utterance. The excessive familiarity that haunts Biron and Armado's insistent expressions of love melancholy comes across as a kind of parody of Judith Butler's model of citational performativity, in which every particular performance of a given identity necessarily recalls and draws intelligibility from a normative background of prior performances:

> Performativity is thus not a singular "act", for it is always a reiteration of a norm or a set of norms [. . .] this act is not primarily theatrical [but] its apparent theatricality is produced to the extent that its historicity remains dissimulated (and conversely, its theatricality gains a certain inevitability given the impossibility of a full disclosure of its historicity). [. . .] Indeed, could it be that the production of the subject as originator of his/her effects is precisely a consequence of this dissimulated citationality?[18]

The performativity of masculine love melancholy deforms (or aggressively hyper-conforms?) Butler's model from within, for the pressure of prior examples and norms is so strongly apparent to both the love melancholic and his audience that that citationality, far from being dissimulated as a condition of the construction of origins, enters the discourse of the melancholic directly as another gambit for self display. Far from denying any relation to prior examples, the masculine love melancholic vocally protests at the suffering of his wounded authenticity beneath the collective weight of all the plackets, codpieces, sighs, and love-rhymes which precede him. This gesture calls attention to itself as a uniquely local site of physical and emotional pain at precisely the same moment that it resoundingly announces itself as a reiteration of a prior discursive norm, the tradition of lovemaking. The goal is to secure in the loved one and/or the audience that suspension which Octave Mannoni formulates as "I know very well, but just the same . . . ."[19] In a manner akin to Jaques' assertion that, despite all the exhausted and widely recognizable social modes of melancholy, his own suffering is nonetheless "a melancholy of mine own," in this passage it is as if Biron is imagining Rosaline's response as: "I know very well that all melancholy is just a courtly fashion, but this time it's different." As Carla Mazzio notes in her chapter on "the melancholy of print" in *The Inarticulate Renaissance*, the frustrated wits attempting to express their own feelings in *Love's Labour's Lost* are constantly pressurized by "the sheer proliferation of amatory discourses [. . .] in a world dominated by technologies of writing and print."[20] Accordingly, melancholic authenticity is approached asymptotically along a curve defined by prior traditions, examples, and predecessors who crowd the moment of utterance: The more polished a given performance of love melancholy is, the more it follows the contours set by the conditions of citational performativity that precede it and make its recognition possible. But as Biron's chest-thumping demonstrates, at the final approach all that is left is a deictic "hard sell" that *this* time, I am for real.

In *The Melancholy of Race*, Anne Cheng responds to Butler's distinction between performativity and performance with a question which articulates from another angle the same ambition we detect in Shakespeare's melancholics, expressing a desire to somehow dislodge the threat of citational determination from within the performance itself:

> Can performance ever disturb performativity? Or does that question retreat into an atavistic idea of agency that a concept like performativity wants to qualify? Surely there must be a play between the two poles, moments when performance outstrips performative constraints and vice versa?[21]

Searching for a notion of agency that will allow for critical engagements with prior norms and perhaps feeling that Butlerian performativity is too claustrophobic, Cheng's questions are infused by a theoretical desire for some kind of leverage against total determination by the prior context of performativity. Cheng finds such leverage in performances marked by a tension between the specificity of the performing body and its contextualizing frames of performativity.[22] Cheng proposes, tentatively, a possibility that the body could speak symptomatically in a voice that evades the snares of citation. While Butler has a response to such a move (indeed, it seems that *Bodies That Matter*, with its deconstruction of the Greek terms "*hyle*" [matter] and "*physis*" [nature], is in part meant to unravel the supposed preconstructivist integrity of "body" as a foundationalist fantasy), I am more interested in how Shakespeare handles such a possibility than in rehearsing Butler's already-ample defenses.[23] *Love's Labour's Lost* seems to propose the female body as, among its many significations, a melancholic space that evades the discursive insecurities and manic overdeterminations of Biron and Armado's melancholic performances. Here we are prompted to imagine melancholy assemblages as "affinity groups" consolidated around ongoing, gendered divisions of affective labor that rest upon an implied but elusive biological poetics of difference between male and female characters.

"You That Way. We This Way.": Melancholy Transmission from Male Minds to Female Bodies

Going beyond the occasional spasms and gestures that punctuate Biron's ravishing verbal set pieces, the play comes to articulate the possibility of kinds of pain and death that are not mere turns of phrase. In the play's final act, after a series of puns about Cupid and pejorative references to him as a boy tyrant which echo Biron's extended tirade, Katherine darkly describes Cupid as a "shrewd unhappy gallows," which prompts Rosaline to reply:

> ROSALINE: You'll ne'er be friends with him; a' kill'd your sister.
> KATHERINE: He made her melancholy, sad, and heavy
> And so she died. Had she been light, like you,
> Of such a merry, nimble, stirring spirit,
> She might ha' been a grandam ere she died.
> And so may you, for a light heart lives long. (5.2.12–18)

In his *Appreciations*, Walter Pater singles out this brief glance toward mortality for particular praise: "[T]he lines in which Katherine describes the blighting through love of her younger sister are one of the most touching things in older literature."[24] Pater's appreciation of the pathos of this briefly sketched incident necessarily ignores the speed with which it is dropped in favor of yet another punning double entendre on "lightness" as a quality of both one's persona and of one's sexual honor. But if the banter continues unabated, it takes on a somewhat more strained valence following Katherine's sobering revelation. The intimacy established by Rosaline's familiarity with the sister's death makes Katherine's rehearsal of it all the more disturbing. Subtly preparing the audience for the eventual arrival of Monsieur Marcadé and the more traumatic interruption of a death in the royal family that concludes the play, this anecdote also radically revises the play's construction of love melancholy. The tale of Katherine's sister offers a kind of "proof" of melancholic authenticity that performs a retroactive work upon the discursive models of masculine love melancholy that precede it, correcting Moth's reduction of melancholy to a matter of appearance by counterexample. If love melancholy can kill people, then melancholy is not just a matter of appearance after all. Furthermore, by contrasting male pretensions toward deep feeling with a more persuasive (however anecdotal) account of female seriousness and steadfastness, this counterexample parcels out melancholy authenticity along gendered lines: Men falsely express a melancholy that they do not feel, while women suffer and die from a "genuine" (feminine) form of love melancholy whose seriousness is not subject to debate.[25]

This anecdote stands in an ambiguous relation to contemporary medical understandings of melancholy. If Armado and Biron's manic performances configure melancholy as discursive, Aristotelian, "light," genial, and masculine, the anecdote of Katherine's sister reconfigures melancholy as somatic, Galenic, "heavy," morbid, and feminine. The notion of "heaviness" which Katherine ascribes to her sister suggests an imbalance consistent with the Galenic medical tradition. As Galen describes it in Book III of *On the Affected Parts*, the pathological form of melancholy consists in an excessive localized production of black bile, leading to an excess of cold and dry.[26] The liver can heat black bile into a toxic form known as melancholy adust, which can then affect the brain, the blood, or the intestines, with each particular zone exhibiting distinct symptoms. The circulation of this additional substance within the bloodstream was thought to lead to the kind of symptomatic "heaviness" which Katherine describes. Where Shakespeare's text diverges from contemporary medical models is in its outcome. Medical theorists who describe melancholy itself as leading to death are extremely rare; melancholy was not thought of typically as a fatal illness in and of itself.[27] The answer to the question of whether Katherine's sister died of melancholy or was driven by her melancholy to commit suicide lies inside a kind of textual blind spot, an encrypted space which conceals as much it reveals, memorializing the dead sister and forgetting her in equal measure.[28] By subtracting the information that could stabilize the event and place it clearly under the categorical control of a particular melancholic model or scenario, Katherine's speech works to increase the trauma of her sister's death. The pathos that Pater and other critics find in this dark episode articulates a desire occasioned by this representational restraint.

It is a critical commonplace to detect in the story of Katherine's sister a "foreshadowing" of the death of the King of France, the event that arrests the sequence of manic entertainment and courtly wooing. To reverse this directionality, one could argue that the "unspeakability" of the death of the King, in its very emptiness of content, merely repeats the absent core of Katherine's sister's death, its lack of "back story" finding an echo in Monsieur Marcadé's broken off report:

> *Enter a Messenger, Monsieur Marcadé.*
>
> MARCADÉ: God save you, madam.
> PRINCESS: Welcome, Marcadé,
> But that thou interrupt'st our merriment.
>
> MARCADÉ: I am sorry, madam, for the news I bring
> Is heavy in my tongue. The king your father—
> PRINCESS: Dead, for my life!
> MARCADÉ: Even so. My tale is told. (5.2.706–712)

The missing copulative "is" that should fall between Marcadé's "The king your father—" and the Princess's answering "Dead" formally registers the fact of death as a gap in representation. The unrepresentability of "being dead," its radical impossibility as a space for thinking, is registered as a catch in Marcadé's throat, a liminal heaviness upon the tongue which stops speech short. Marcadé functions as a kind of marker (one who "mars" the "arcadia" of Navarre) who has only to show up and begin to name the father in order to suspend enjoyment; in total opposition to the kind of sumptuous verbal pictures of death which Shakespeare occasionally allows himself (of which Gertrude's impossible account of Ophelia's drowning is the most notorious), here one has only to name the principal character and "the tale is told."[29] Caught in the midst of "merriment," the heaviness which Marcadé (and by extension the court and the audience) experiences symptomatically repeats the original melancholic heaviness of Katherine's unnamed sister. In a sense, this announcement also encrypts the death of the father, concealing its scenario from view even as it renders it public, shareable, and open.

Such a model strikingly resembles Nicolas Abraham's account of "the Somatic" and provides us with a psychoanalytic model for how to think through the representational challenges occasioned by the bodily "interior":

> The Somatic must be something quite different from the body proper, which derives from the psychic as one of its functions, the psychic having been described by Freud as an exterior layer, an envelope. The Somatic is what I cannot touch directly, either as my integument and its internal prolongations or as my psyche, the latter given to the consciousness of self; the somatic is that of which I would know nothing if its representative, my fantasy, were not there to send me back to it, its source as it were and ultimate justification. The Somatic must therefore reign in a radical non-presence behind the Envelope where all phenomena accessible to us unfold. It is the Somatic which dispatches its messengers to the Envelope, exciting it from the very place the latter conceals.[30]

Mr. Marcadé, messenger from the Kingdom of the Somatic, brings death to the discursive revelry of Navarre. As signaled by its missing yet implied copula, death is and is not present, manifesting itself only by way of a messenger, and so remaining outside, something we cannot touch directly, but whose tokens in fantasy and narrative (in announcements that the king is dead, in stories of a sister who died of melancholy) condition our access as tantalizingly partial, incomplete yet undeniable. The intrusion of the Somatic into the manic rhythm of the play's entertainments is sudden, disorienting, decisive. The dramatic experience of interruption is shocking and forces a revision of both the courtly entertainments and amorous farces that precede it. What was merely "light" now lacks excuse and triggers shame, even as its enforced blockage retroactively reinforces a certain sweetness and gentility.

These deaths are unanswerable. They stop the mouth in the midst of speech, drain and dissolve the energy required by the country dance, and replace the light heart of a comic courtier with the heaviness of tragic mourning. What remains of the play is dedicated to the painful task of learning how to channel the manic energies that sought to avoid any consciousness of limits now that such knowledge has intruded. By grinding down distinctions of status, authenticity, and identity in a universal cull, death poses the ultimate threat to the self-limning project of masculine melancholic display that Armado and Biron have engaged in under the broader project of fashioning a "little academe." The homosocial community that Navarre sought to construct (mockingly referred to by Boyet as "the Prince and his bookmates"), having already been soundly beaten at its own game of wits by the visiting Princess and her ladies, is now harrowed further by the enforced silence and separation which the death of the King of France imposes (4.1.96). From the collective satirical mockery of the pseudo-melancholic outcast to the solidarity of mourning for its genuine victims, melancholy offers the opportunity for the cohesion of social assemblages. From both ends, melancholy cuts some individuals off in order to join others together upon the site of that very absence. *Love's Labour's Lost* commences an exploration of these generative disturbances within melancholy, with Shakespeare raiding its richly contradictory heritage of medical and social knowledge only to stop short before the unknowable somatic abysses which melancholy also occasions. It is in the *Merchant of Venice* that this dynamic of masculine display and bodily limits upon knowledge is pressed further. In the jump from Navarre to Venice, the circulation of melancholy within the social body becomes both more productive and more virulent.

3. *Let Me Have Judgment, and the Jew His Will*

Weariness and Wariness

Foreclosing knowledge from its first line, *The Merchant of Venice* may begin but it does not quite open. We start out startled, at impasse, greeted by this concession of defeat:

> ANTONIO: In sooth, I know not why I am so sad.
> It wearies me, you say it wearies you;
> But how I caught it, found it, or came by it,
> What stuff 'tis made of, whereof it is born,
> I am to learn;
> And such a want-wit sadness makes of me
> That I have much ado to know myself. (1.1.1–7).[1]

For Antonio, melancholy is not only an illness—it is a discourse, a field of study, and one whose daunting curriculum (etiology, source, transmission, substance, and origin) mocks his exhaustion. Presented with this failure in the play's first line, *The Merchant of Venice* commences with the provocation to interpret melancholy pitched oddly between opportunity, challenge, and therapeutic responsibility. Its fundamentally epistemological mystery anticipates the seeming causelessness central to Burton's portable definition of atrabilious disease in *The Anatomy of Melancholy* (1621): "a kind of dotage without a feaver, having for his ordinary companions, fear and sadnesse, *without any apparent occasion*."[2] Regardless of whether we take Antonio's negation of self-knowledge as a direct confession of a problem he cannot solve or as a coy evasion of a truth he'd rather not speak, his statement is a conversational gambit: It constitutes his sadness as an interpretive puzzle for others, within and without the play. His melancholy at once creates and resists knowledge.

We already know what psychoanalysis would have to say about this condition: the translation of early modern "melancholy" into Freudian "melancholia." The widespread adoption and absorption of psychoanalytic terms and phrases into ready-to-hand critical commonplaces has of course been curiously simultaneous with the equally prevalent and canonical acknowledgments of its reductions and biases, the bagging and tagging of the historical, cultural, and hermeneutic horizons of what Derrida was already calling in 1979 "these extremely old matters."[3] In the case of Antonio—and he is presented and framed for us in this scene as a "case" complete with diagnostic speculation and a therapeutic race for a cure—one might expect psychoanalytic theory to encourage us to dutifully uncover the lost object of love, the person or idea that has undergone a symbolic "death" and which is now mourned and encrypted beneath the troubled surface of the melancholy symptom.[4] Registering the overfamiliarity of such a reading, in a recent essay on the play Henry S. Turner speaks for the impatience of many in politely wanting to change the subject:

> "How are we to explain the cause of Antonio's sadness? One ready psychoanalytic answer—a melancholia brought on by the failure to incorporate a lost Other; a generalized mourning constituted by a lack in the subject's (unconscious) refusal of a (homo)erotic object—risks foreclosing full considerations of the play's ethical and political difficulties."[5]

The readiness is all. Rushing to defense, one could simply protest that it is not psychoanalysis that does the foreclosing, but a too pat reduction of the psychoanalytic interpretation of melancholia into an Easter egg hunt for a lost Object, in which such losses are taken to be necessarily prior to the ethical or political constitution of the subject. I do not see that such an orientation would necessarily exclude or preclude those domains (i.e., despairing thoughts of personal damnation could prompt the loss of the self as an object of love for ethical reasons; internalized shame at one's exclusion from citizenship could prompt the loss of the self as an object of love for political reasons, etc.). That is, I take even a reductive account of psychoanalytic reading to at least potentially assist us in considering the ethical and political difficulties rather than to foreclose or preempt them, and I intend in this chapter to do just that. But one must also take his point as an index (dare one say symptom?) of a broader phenomenon: the waning critical purchase of psychoanalytic explanations within Shakespeare studies.

One would not expect a critic who is simply, as they say, "of another party" to tarry with psychoanalytic terms and concepts while getting on with the business of interpreting Antonio's mysterious affect within *The Merchant of Venice*, but the absence of such critical rubrics is more striking in recent interpretations of the character and the play from psychoanalytic critics themselves. Janet Adelman and Julia Lupton are exemplary here. Their recent accounts of Antonio within the play (Lupton's chapter on *The Merchant of Venice* in *Citizen-Saints* and Adelman's *Blood Relations: Christian and Jew in the Merchant of Venice*) push off from psychoanalytic language and frameworks and instead concentrate upon intellectual and religious history, political theology, and scriptural exegesis.[6] Thus the weariness of Antonio in thinking about the cause of his own sadness finds its mirror and complement in a certain ambient wariness on the part of early modern literary critics about the explanatory rewards of psychoanalysis when attempting to understand Antonio. Case closed?

I share the wariness, but I beg to differ. The potential oversimplification of a hostile takeover that the standard theory of melancholia offers up when brought to bear upon Antonio are not signs of the limits of psychoanalysis itself, but of a too-credulous reliance upon the translation of early modern melancholy into its clinical near homonym. The interpretive pathway into Antonio's self-defeating sadness that psychoanalysis offers us should not proceed along the standard route toward its preordained answer, but ought to tackle the antagonism and difficulty of this character, the resistance to knowledge that hides behind the demand to be known. How do we connect Antonio's desire for the source of his sadness to be known with his curious, indeed queer, desire to be seen to suffer that threatens to erupt in the trial scene?

If Antonio begins the play as a melancholic who cannot name the source of his despair, by the trial scene he has found a way out: the imminent spectacle of his own torture and death. Antonio's anxious cry "let me have judgment, and the Jew his will" (4.1.83) articulates a desire that is only explicable if we go beyond the clinical understanding of melancholia and instead consider its contours in terms of an altogether distinct—but no less psychoanalytic—critical formulation: masochism.[7] Retroactively, the contract between Antonio and Shylock can itself be understood as a classic masochistic contract, and expresses a collaborative dynamic of suffering and spectatorship that we already see underlying Antonio's initial display of melancholic suffering. Were this account to simply burrow into Antonio's psyche in search of further local confirmation for a freestanding logic of the symptom, the result would entirely grant Turner's scenario of foreclosure: The psychoanalytic reading would indeed disavow the ethical and the political stakes, privatizing critique at the level of the literary subject. But as I hope to show, the structure of subjection and desire at work "within" Antonio's melancholy and his masochism bears a wider family resemblance with the buried contradictions that drive the turning worlds of Venice and Belmont. The extraction of the pound of flesh is not Antonio's masochistic fantasy alone but the political and ethical fantasy of subjection at the heart of the social assemblage that surrounds him.

The "Standing Pool" of Melancholy

Antonio's opening assertion of nonknowledge does not go unchallenged. In confidently suggesting that Antonio's sadness is not a mystery at all, but simply comes as a result of the financial insecurity of his risky business ventures, Salerio and Solanio make a double introduction, conveying both Antonio's occupation as the titular merchant and sketching the dangers and rewards of the broader world of Venetian mercantile capital. Salerio's account of how the visual details of one's surroundings could call up associations with the absent ships laden with merchandise provides a miniature account of the workings of the melancholy mind:

SALERIO: Should I go to church
And see the holy edifice of stone
And not bethink me straight of dangerous rocks
Which, touching but my gentle vessel's side,
Would scatter all her spices on the stream,
Enrobe the roaring waters with my silks,

And, in a word, but even now worth this,
And now worth nothing? Shall I have the thought
To think on this, and shall I lack the thought
That such a thing bechanced would make me sad? (1.1.29–38)

If Antonio initially depicts the melancholy subject as marked by an aporetic void or resistant cluster of nonknowledge somehow within the subject, Salerio's remark rethinks this, supplementing Antonio's definition of melancholy as a kind of dumbfounded ignorance with a compensatory potency, familiar from the tradition of pseudo-Aristotelian genial melancholy: the claim that melancholy provides the subject with access to truth.[8] Melancholic speculation upon capital can articulate a kind of hidden knowledge of negation that is always threateningly present within it, if only as a latent possibility. In the play's opening lines, then, we have shifted from a portrayal of the melancholic as the one who does not know to a portrayal of the melancholic as the one who knows all too well, who attends too closely to an unanswerable threat which is both universally, trivially true (the commodity can always be destroyed) and fundamentally "unhealthy" to think about (i.e., hostile to the capitalist narrative of profit and increase and thus in a sense, literally counterproductive). To dwell too attentively upon such a possible outcome is, to borrow a phrase from Horatio, "to consider too curiously" (*Hamlet*, 5.1.192). The stakes of this counterproductivity of melancholy knowledge will become considerably amplified in Antonio's eventual identification with the "tainted wether of the flock," but what is interesting at this point is that this model of melancholy knowledge is precisely not claimed or affirmed by Antonio himself but produced by Salerio spontaneously and projected onto Antonio as a possible content for his troubling emptiness. The very constitutive blankness of Antonio's melancholy, far from stopping discourse short, thus seduces and triggers it. Salerio's image of a loaded merchant vessel suddenly bursting open and revealing its contents figures the task of melancholy interpretation itself; the rock pierces the sturdy side of the ship, penetrating the valued content within, making it available yet also radically emptying it of value. In an anticipation of Antonio's later shift from melancholic to masochist, "being known," "being opened," and "being destroyed" are all brought into a charged proximity from the very beginning of the play.[9]

With a keen therapeutic thrust, Graziano interrogates the desire behind Antonio's announcement of nonknowledge and issues the first direct challenge to Antonio's melancholy gambit, suggesting that melancholy is not a retreat from the social but simply another way of circulating within it:

> GRAZIANO: There are a sort of men whose visages
> Do cream and mantle like a standing pond,
> And do a willful stillness entertain
> With purpose to be dressed in an opinion
> Of wisdom, gravity, profound conceit,
> As who should say "I am Sir Oracle,
> And when I ope my lips let no dog bark!" (1.1.88–94)

Graziano's caricature of the somber melancholic is so humorous and distracting one almost does not notice that in fact Antonio's own melancholy strives to achieve precisely the opposite effect.[10] Far from seeking to induce a silent awe and respect, Antonio strives to generate conversational interest in his secret, repeatedly drawing those around him into its analysis. The effect of Antonio's melancholy is not to deepen the level of discourse progressively but to hold stubbornly at the surface. Far from oracular, as Antonio entertains and rejects his companion's diagnostic sallies, he seems flatly chatty but blank, at least until Bassanio arrives on the scene. But while Graziano's speech does not "hit home," it does reinscribe the play's concern with the epistemology and poetics of opacity: Does the creamy standing pool conceal hidden treasure in its depths, brackish wastewater or simply more surface?

Rhetorical play upon the lack of correspondence between interior nature and exterior show is, of course, the melancholic gambit par excellence. As a trope of melancholic identification, it hints at an inward richness of feeling concealed by the stony external posture of melancholic stasis; as a trope of critique, it allows the melancholic to attack others aggressively and suggest that their surface goodness and happiness mask a hidden reverse of shame, guilt, and morbidity. Shakespeare repeatedly places such imagery of deceptive surface and hidden truth in the mouths of melancholics, culminating with Hamlet's obsessive images of the morbidity and infection concealed beneath the bodily surface,[11] and there are many analogues to this dynamic within *The Merchant of Venice*, from the gold casket scene with the Prince of Morocco to Antonio's proverbial excursus:

> The devil can cite Scripture to his purpose.
> An evil soul producing holy witness
> Is like a villain with a smiling cheek,
> A goodly apple rotten at the heart.
> O what a goodly outside falsehood hath! (1.3.96–100)

While it shares this standard surface/depth topology, Graziano's use of the trope critically transvalues its standard articulation. Concluding his sermon, Graziano asks that Antonio "fish not with this melancholy bait / For this fool gudgeon, this opinion" (1.1.101–102). Antonio stands accused of cynically exploiting the very difference between surface and depth which structures the melancholy trope. This discursive reduction of melancholy is here left hanging, unanswered and perhaps unanswerable. This image of fishing with melancholy bait interlocks with the preceding image of the standing pool, with the blankly serious face of melancholy functioning as both pool in which one fishes (for the meaning of sadness) and bait with which one fishes (for the respect and opinion of others). Graziano's mixed metaphor is at once a comic failure of wit and itself the commencement of a subtly violent poetic in which the epistemological quest for a fixed content for melancholy is articulated through fantasies of opening the body. Here physical opacity codes epistemological barriers, but it is not simply that frustration at the latter triggers fantasies of violence upon the former. Rather, the very stasis of melancholy, its static fixity and physical passivity are somehow represented as an affront, inciting violence. The creamy surface skin cries out to be ripped or slit open to reveal the supposed truths concealed beneath its ostensible blankness, a blankness which incites a violent anatomization of its interiority because of its "too, too solid" material status. The standing pool of the melancholic face is deceptive insofar as a solid skin has formed over a liquid pool—the very solidity of the "creamy mantle" represents not just enclosure but deception, treachery, masquerade.

The play repeatedly figures the violation of Antonio's melancholic body with such imagery of fishing, and not always in Graziano's gently comic tone; later in the play this poetic recurs chillingly in Shylock's reaction at the news that Antonio's ships have foundered:

SALERIO: Why I am sure if he forfeit, thou wilt not take his flesh—what's that good for?
SHYLOCK: To bait fish, withal. (3.1.48–50)

Here the extraction of the pound of flesh threatens to gruesomely literalize Graziano's metaphor of "fishing with melancholy bait" as Shylock imagines himself baiting fish hooks with the morcellated pieces of Antonio's melancholy body. Far more than just an index of his "Jewish" cruelty, this remark must be read at several levels, both in terms of Shylock's own character and in terms of the poetics of the play as a whole. At the most basic level of the received wisdom which is depicted as ambient within the world of the play, Shylock's remark echoes and confirms Graziano's prior insight that the melancholic uses his body as bait, dangling it before others, teasing them with the possibility of knowing his interior secret. The stock melancholic expression of puzzling sadness or the public announcement that "I know not why I am so sad" are also forms of bait for spectators, tools one uses to acquire the cultural capital thought to attend the privileged status of the melancholic as oracular sage or genius: This is the core idea that unites Graziano's first image of the standing pool to his secondary accusation of fishing with melancholy bait. But Shylock's final echoing repetition of the same fishing image sees that accusation and raises it by insisting upon a literal enactment of what was for Graziano only a turn of phrase.

The play also makes available an anti-Semitic reading of this remark which would detect in it the workings of a certain "Jewish" literalism (Shylock voices not a metaphor about fishing but an actual plan to fish) and "Jewish" thrift (as Shylock sees it, even the dead can be put to work). Within this repetition, there is perhaps a marking of a "Jewish" difference that makes a difference: Shylock's imagination contrives an economically productive use for what his Christian interlocutor Salerio regards as a worthless, excremental remainder, providing a further example for the play's anti-Semitic spectators of the seemingly magical capacity of Jews to extract a surplus of capital from dead matter. Where a Christian sees an end to circulation, the Jew is here shown to discover the possibility of one more transaction through a kind of funereal usury. Bracketing the security with which we can correlate such an insight with the play's complex construction of "Jewishness," there is a basic work of confirmation and reinforcement here, for Shylock's remark ironically ratifies Graziano's insight that the melancholic remains in circulation and sutures together the epistemological and economic valences of the fishing pool image. It does so by translating an economy of cultural capital (where respect and esteem are gained or lost by attempts to trick others into belief) back into a material economy in which beliefs are bracketed out of the picture and value is not a matter of rhetorical persuasion but material possession. What is refused by Shylock in this work of translation (and it is a blind spot which will haunt him in the courtroom) is precisely the mirage that hypnotizes both the melancholic subject and his Christian spectators: the possibility of absolute loss.

When we consider the consolation scenes that begin the play as a sequence (for surely Portia's complaint that "my little body is aweary of this great world" [1.2.1–2] is a melancholic echo of Antonio), Shakespeare seems at pains to present an inconclusive series of distinct, mutually contradictory readings of melancholy: as a problem of knowledge, as a sound but comic reaction to financial pressure, as a calculated pose, as an economically based surfeit.[12] All of these possible answers have an intuitive appeal, and yet all are partial, and none seems quite sufficient. This solicitation of our interpretive energy combined with a foreclosure of any possible solution constitutes the epistemology of melancholy as a double bind: a proliferation of potential causal theories which overdetermine the reason for melancholy on the one hand, and a poverty of verifiability about the inner experience being attested to on the other hand.

The effect of this proliferation is not just a confirmation of ambient sixteenth-century skepticism, though that is one of its effects. This dynamic of drought-in-glut generates a kind of textual inflammation: a hovering, nonspecific curiosity, a dramatic heightening of desire for access to an absent interior, a hybrid imaginative space between Antonio's self or soul and his body. This interior can be understood in psychological terms as a kind of psychic self-encryption of affect, and speeches such as Antonio's opening lament about self-knowledge suggest a working definition of melancholy interiority in terms of mental privacy, or the self-presence of a thinking subject.[13] But this interiority can also be literally "fleshed out" as a kind of somatic state or mode of melancholic embodiment in which the body is subject to a humoral excess of melancholy adust and is flooded with a particular fluid.[14] This second understanding, of melancholic interiority as a physically inner space that could actually be opened up, is also imagined in images of formal anatomization and corporeal destruction, such as Salerio's description of the breached ship's hull spilling its contents. It is this ambition to be known through a violation of interiority that the melancholic solicits, and that the extraction of the pound of flesh threatens to realize so gruesomely.

Biting at the melancholy bait, countless critics and audience members have responded to the text by proposing various definitive answers to the question of Antonio's sadness, thereby entering the assemblage of comforters and would-be eulogizers which his melancholy suffering and contractual doom has called forth.[15] Such a reaction misses the extent to which, as Graziano suggests, melancholy operates within the play by hovering above the grounding that any particular answer would provide. Once his ships have foundered and he is caught within a contract that mandates his physical suffering, the flexible vagueness or productive "blankness" of Antonio's melancholy adapts accordingly, downshifting from the discursive to the somatic. Antonio's initial desire to know himself and to be known by others, the melancholy gambit which commences the play and guides his interactions with the men of Venice, quickly modulates into a desire to be seen, and to be seen to be suffering physical pain. The desire to be seen is a revision of the desire to be known that promises to resolve at the level of fantasy Antonio's melancholy problematic. This translation of the play's poetics of "opening" from the epistemological to the visual register determines Antonio's conditional embrace of his sentence:

> ANTONIO: These griefs and losses have so bated me
> That I shall hardly spare a pound of flesh.
> Tomorrow, to my bloody creditor.
> Well, jailer, on. Pray God Bassanio come
> To see me pay his debt, and then I care not. (3.3.32–36)

Facing the threat of Shylock's bond with a provocatively indeterminate mixture of heroic resignation and self-destructive longing, Antonio's "standing pool" now offers a scenic point of speculation into the motives beneath its surface calm. The same problems of identity and intersubjective knowledge which obtained in the debates about the cause of Antonio's sadness in the play's opening act are sustained and intensified by his troubling reaction to the mortal gravity of the bond. When literary critics and audience members struggle to determine a motivation for this weirdly passive stance, the same promiscuous abundance of explanatory frameworks is called forth only to fall short of catching any definitive hold. Does he act out of love or hate? And love and hate for whom? The task of discrimination between the claims of seemingly exclusive discourses (the economic, the religious, the sexual) ends in aporetic defeat, precisely because the play as a whole represents those domains as relentlessly interpenetrated, always in circulation, and circularly defined in terms borrowed from one another.

Antonio's melancholy functions as a discursive switch point that allows it to "carry" any and/or all of the multiple, overdetermining explanations which his behavior solicits: merchant capitalist anxiety, Christian heroism, unrequited homoerotic desire, moral masochism.[16] Constituted by its affective repertoire of stances, poses, and symptoms, and the recognition they enable, melancholy can function without loyalty to any of these local categories but draws opportunistically upon all of them. Such an ability to be both everywhere and nowhere has been modeled compellingly in Richard Halpern's account in *Shakespeare Among the Moderns* of "the Jew" as systematized by anti-Semitic fantasy. Drawing upon Marx's analysis of "the Money-form," Halpern's analogical argument argues that "[t]he Jew is forced to play a representational role with respect to bourgeois cultural values very like the one that money plays with respect to economic value."[17] Halpern describes the paranoid, additive logic through which the identity of "the Jew" is made to accommodate seemingly contradictory ideological valences, a representational overload which he graphs as an example of "general equivalence" in which Freemason = Jew, Jacobin = Jew, Bolshevik = Jew, Capitalist = Jew, and so on, endlessly: "In this static, paranoid structure, the Jews are not part of a general flow or exchange but are the one term capable of connecting others (such as Bolsheviks or bankers) who would otherwise seem antithetical."[18] The figure of the melancholic proves equally susceptible to such a proliferation of multiple, potentially antithetical identifications in which Capitalist = Melancholic, Christian martyr = Melancholic, Homosexual = Melancholic, Masochist = Melancholic, and so on. The list of potential additional identities is long; I have sketched here only those the play seems to propose as possible roles for the "sad part" which Antonio is playing. Halpern's structure provides a model for thinking through overdetermination which is not, as it were, a lateral profusion of equally likely causes but a kind of bottleneck in which multiple, equally "in-play" components of lived social subjection (to norms of sexuality, to economic practices, to civil laws) are routed through a particular, organizing trait whose particular qualities allow a kind of suspension between conflicting, mutually incompatible rationales.

In the opening consolation scenes, the play insistently locates melancholy identification in relation to economic determination, as either the consequence of the risks of merchant capital or the emptiness imagined to attend its acquisition. But the play also keeps open the possibility of melancholy as the consequence of repressed sexual desire or barred romantic longing. The constitutive blankness of the melancholic's "that is not it" in relation to any proposed solution to the mystery of their sadness forecloses the possibility of adjudicating between these competing rhetorics, and keeps the engines of interpretation and attention turning over.[19] This is particularly the case with the curious symmetry in place between melancholy and homoerotic desire as simultaneously available/resistant explanatory rationales for understanding the affect that underwrites the bond in the play between Antonio and Bassanio.

The mixed plenitude of possibilities and poverty of determining evidence about the cause of Antonio's sadness constitutes his challenge to the community that surrounds him and the audience members that watch him. Such a stance threatens to turn any critical response into a merely symptomatic repetition of the melancholic guessing game of the play's first scene, in some sense forcing the reader to take Antonio's mysterious sadness at its word in the very moment of attempting to sound the depth of his "standing pool." And yet, to borrow a Heideggerean locution, that is precisely how Antonio shows up for us, as audience members, as a problem to be solved. But if in his attempts to relate himself to himself, Antonio is behaving as a thoroughly conventional stage melancholic, his interactions with the principal characters of the play (Shylock, Bassanio, Portia) require another logic in order to be explained. Antonio's knotting together of public hatred for Shylock, self-sacrificial love for Bassanio, and competitive symmetry with Portia can be best understood not in terms of Freud's remarks on melancholia but in terms of the Freudian paradigm of masochism. Antonio's melancholy both grounds and draws retroactive support from the masochistic fantasy that he enters into through his conditional embrace of his juridico-legal fate at the hands of Shylock. The transition in the characterization of Antonio within the play, which I have loosely described as a shift from melancholy to masochism, is thus not a repudiation of melancholy but its Hegelian *Aufhebung*, a sublation that preserves what it also cancels. Antonio's sense of himself as a melancholic "playing his sad part" prepares him for his role in the masochistic fantasy constructed by the play.

Antonio as the Subject of Masochistic Fantasy

To avoid reductive overfamiliarities, it is important to draw from the start a distinction between reading Antonio in particular as a masochist and reading the plot arc bounded by the contract scene and the trial scene within *The Merchant of Venice* together as scenarios within a masochistic fantasy. While it may be that the reading of the latter requires to some extent the kinds of identifications and speculations typical of the former, my goal is not only to offer an analysis of a particular character but also to make a broader claim about the relationship between text and reader in which fantasies of mastery and subjection, contract-making and cruelty are both generated and dissolved. Building affective assemblages across and between bodies, the melancholic recruits interpreters and the masochist conscripts torturers, with each staging their identity for others by forcing everyday social relations to ascend to the level of insistent, willfully "excessive" scenarios of suffering or punishment, scenarios which seem to call forth interpretive labor or collaboration.[20] These forms or subjects of discourse (the melancholic, the masochist) provide a pliant template around which the various communities of the play can organize themselves as comforters, as rescuers, as judges, as friends.

That said, the designation of the trial scene as the staging of a masochistic fantasy prompts the question of how we are to define both "masochism" and "fantasy," and how they can be deployed outside the clinical context of their psychoanalytic formulation. In *The Language of Psychoanalysis*, Laplanche and Pontalis define fantasy as an "imaginary scene in which the subject is a protagonist, representing the fulfillment of a wish in a manner that is more or less distorted by defensive processes."[21] In a useful supplement to this basic definition, in her essay "Fantasy: An Attempt to Define Its Structure and Operation," Maria Torok emphasizes the experience of fantasy, which she describes as an intrusion from the imagination into consciousness of material which she terms a *misfit* with its local context, which calls attention to itself as intrusive, which seems somehow both "wrong" and insistent.[22] Moving from the clinic to the playhouse, the audience's experience of Antonio's fantasy conforms strikingly to Torok's definition, for it is both counterintuitive and conspicuous, in need of some explanation, and at odds with any rational instinct of self-preservation. Faced with a terrifying ordeal, Antonio seems to advance the claim, "I do not object to being tortured (in fact I desire it) so long as I am watched by him." Antonio tolerates the threatening imminence of his own torture and death by imagining that torture transformed into a spectacle for another, a ritual staged for an audience. In her final account of how fantasy acts upon and within relationships of transference, Torok suggests that the intrusion of fantasy articulates just such a collaborative impulse: "[F]antasy constitutes a positive attempt to transcend the pure affect and arrive at a representation of it, an attempt in which the analyst is invited to participate. [. . .] Fantasy is expressive of an attempt at working through the problem and is combined with a desire for collaboration."[23] Antonio's active solicitation of others (as interpreters, as partners in the "problem" of his sadness, as witnesses of his suffering) parallels this transferential plea for participation that Torok detects. The conditional acceptance of Shylock's bond, so long as Bassanio will appear as "star witness," shares this structure; Antonio's fantasy requires the partnership of others as one of its conditions of satisfaction, and already casts them within it from its first articulation. His letter to Bassanio at Belmont frames this collaboration between the two in terms of his own desire to see his friend: "[A]ll debts are cleared between you and I *if I might but see you at my death*" (3.2.318; italics mine). But the precise valence of this conditional acceptance of his suffering and death is dramatically reversed when Antonio cries out at the close of the next scene: "[P]ray God Bassanio come / *To see me* pay his debt, and then I care not" (3.3.35–36; italics mine). Following from Laplanche and Pontalis' account of the revisions and concealments of the fantastic scene, we can understand Antonio's initial request ("I wish to see you before I am tortured") as a defensive distortion of the fantasy's true valence: "I wish to be seen by you while I am being tortured."[24] As an internally consistent scenario constructed from a plural set of agents, participants, objects, and signs, Antonio's masochistic fantasy constitutes a polychronic assemblage: a virtual future outcome with symptomatic effects upon the present.

Like *fantasy*, *masochism* is a term whose reductive familiarity threatens to do as much harm as good when critically invoked. The creation of the complex manifold of desires, behaviors, and roles known as "masochism" commences in 1886 with Krafft-Ebbing reading, and reading into, literature: "I feel justified in calling this sexual anomaly 'Masochism' because the author Sacher-Masoch frequently made this perversion, which up to his time was quite unknown to the scientific world as such, the substratum of his writings."[25] The importance of Masoch's precipitation of masochism is further borne out by the number of Krafft-Ebbing's patients who cite his novels as important crystallizations of their own erotic ideals.[26] There is a knotty circularity to the relationship between literary example, pathological identity, and personally experienced longing which Krafft-Ebbing's use of Sacher-Masoch commences and psychoanalysis inherits. In her essay "On Masochism: A Contribution to the History of a Phantasy and Its Theory," Gertrud Lenzer articulates this problem: "To Krafft-Ebbing, the content of these fantasies and the actions sometimes connected with them constituted masochism, and masochism, in turn, was made up of these fantasies and practices. In other words, the symptoms of the disease constituted the disease itself."[27] Lenzer's suspicion that there is no underlying content to masochism aside from a kind of checklist of textual components simultaneously simplifies and troubles the currency between psychoanalytic knowledge and its literary "example."[28] Susceptibility to reductive definition unites "masochism" and "melancholy" as social phenomena which can—sometimes—be recognized "in the blink of an eye"; they are both superficially easy to detect, yet difficult to define in any satisfying manner.

Obviously, the mere presence within a narrative of pain and punishment does not in itself entail the operation of masochistic fantasy, for if that were the case then the concept would be so broad as to be meaningless; what is required is that this punishment generate a surplus of enjoyment or pleasure, and typically this translation of pain into pleasure is regulated through the impersonal workings of a contract or agreement that governs the relations between master and slave. Much of the peculiar energy of Sacher-Masoch's novels is expended in the obsessive articulation of the details of this contract, whose impersonal legalism stands in a productive tension with its regulation of the most arbitrary and perverse erotic caprices. This resonant narrative pattern unites Sacher-Masoch's novels and Shakespeare's play. *The Merchant of Venice* is, among other things, a play about a written contract which entails a ritualized spectacle of physical punishment as its outcome, and the slow process through which the bondsman of that contract comes to articulate his desire for that punishment. The contract not only provides Antonio with a role as bondsman, scripting and fixing his position in the masochistic fantasy—it also makes possible the play's dramatic tension, its effect of suspense, a term which is, of course, one of the constitutive elements of masochistic sexuality.[29] By posing a threat, the contract makes a certain masochistic pleasure in the hypothetical possibility of that threat available for Antonio to enjoy. Reading backward from Antonio's acceptance of his fate to the initial sealing of the bond, we can read the contract scene itself as an already erotically charged exchange of power between men. Such a reading is already imagined when Krafft-Ebbing explicitly compares the power dynamic of masochistic relationships to economic relationships such as the one that the contract creates. In cases of what he terms *sexual bondage*, "the will of the ruling person dominates that of the person in subjection, just as the master's does that of his bondsmen."[30] Krafft-Ebbing's "just as" seems to simply assert a comparative analogy between master/bondsman relations and master/slave relations, but in fact it marks a constitutive jump from everyday surface to sexual "substratum" that the discovery of masochistic fantasy beneath the guise of everyday social relations both requires and renders evanescent. It is not that everyday economic relations are here shown to be always already inherently masochistic. Rather, the domination by another's will which is simply taken for granted in everyday economic relations becomes for the masochist a specific source of pleasure, and so a dimension of the everyday is no longer simply endured but is enjoyed.

Suturing Krafft-Ebbing's comparative terms together, Antonio is literally both a masochist and a bondsman, for his pleasure in subjection as bondsman occasions his masochism. From Antonio's position, what was for Krafft-Ebbing a rhetorical figure—the "just as"—becomes a literal equation, which one might phrase thusly: Exactly where I am simply and coldly treated as a bondsman by a master, it is there that I am most clearly enjoying masochistic pleasure. Once each party's obligations toward the other have been undersigned, the execution of duties and the sufferance of penalties becomes "impersonal," simply a function of the organizing text. In this sense, the bondsman stands in thrall to the literal text of the bond rather than to the master who is its representative on earth and carries out its dictates. The masochistic insistence that the contract be carried out "to the letter" (a staple in Sacher-Masoch's plotlines and in the erotic literature of masochism) demonstrates the constitutive importance of the contract, its seemingly magical capacity to drive the narrative, a capacity which will undergo ironic transformation in Portia's wielding of "the letter of the law" in the trial scene.[31]

The interlocking structure of Shylock's and Antonio's complementary hatreds and desires defines the operation of masochistic fantasy in the play: While the contract is entered into as a result of Antonio's self-sacrificial "charity" toward Bassanio, it is Shylock who transforms that contract and exceeds the purely monetary terms of the economic by proposing the pound of flesh as a penalty: "let the forfeit / Be nominated for an equal pound / Of your fair flesh, to be cut off and taken / In what part of your body pleaseth me" (1.3.147–150). Shylock's appraisal of Antonio—which begins in a pun upon his being "a good man" in both credit and morality—ends in an assertion that his body will be used for Shylock's pleasure. I believe we can hear a kind of desire at work in Shylock's discourse about Antonio's body and its uses, a desire which registers more as hunger than as sexual lust:

> *SALERIO:* Thou wilt not take his flesh, what's that good for?
> *SHYLOCK:* To bait fish withal. If it will feed nothing else, it will feed my revenge. (3.1.47–50)

These lines, and others like them ("I will have the heart of him" [3.1.119], "Your worship was the last man in our mouths" [1.3.57]) articulate a destructively violent desire to consume the other, depicting Shylock's lust for revenge as a kind of hunger of the soul, the appetitive inversion of Portia's surfeit and Antonio's exhaustion. Shylock answers Antonio's passivity with a feared (and longed for) aggression, an animal longing that penetrates surfaces and, to borrow Aragon's turn of phrase, "pries [. . .] to th'interior" (2.9.28). Such a mirroring relationship determines our reception of both parties, for the anatomizing interpretive energies which

Antonio's melancholy draws upon itself initially will, of course, later rebound upon Shylock, having been transformed into speculations upon the hardness of his "Jewish heart."[32]

The violence of Shylock's unexpected desire for Antonio's "fair flesh" intrudes upon the comic surface of the play's discourse with the inexplicable force of a fantasy. So far, the play has suggested that affects and emotions (melancholic despair, sexual love) are at the mercy of economic determination; the contract scene strikingly revises this one-way directionality by suggesting that the economic sphere is itself a medium through which extra-economic desires and drives act themselves out. Contracts make the satisfaction of desire possible, but they also enable the disavowal of that desire under the protective shade of their impersonal formality. Beneath its neutral language, the contract functions as a site for a plethora of libidinal investments, generating the promise of masochistic satisfaction for Antonio, and of a correlative sadistic satisfaction for Shylock, under the cover of "business as usual." If Antonio is recruiting Bassanio and Shylock as collaborative participants in his masochistic fantasy, Shylock's own increasingly shrill insistence on the satisfaction of the bond, crowned by his refusal to accept a profitable financial restitution when it is offered, suggests a correlative sadism on his part.[33]

What to make of this complementarity? Freud's account of masochism resonates in complex ways with both Antonio and the will-to-torture within the play that surrounds him. But lest we risk oversimplification, it must be pointed out that Freud's theory of masochism is not static, but shifts dramatically over time from its founding hypothesis that masochism is a redirection of hostility. In the transition from "A Child is Being Beaten" (1919) to the "The Economic Problem of Masochism" (1924), Freud altered his account of masochism substantially, replacing the theory of redirected aggression with the more controversial claim that there is a basic, "biological and constitutional" capacity for pleasure-in-pain entailed by the death drive's teleological flow from excitation toward "the stability of the inorganic state."[34] The result in the later theory is a trinitarian taxonomy of masochism in three distinct but related forms: erotogenic, feminine, and moral.[35] Erotogenic masochism provides the basic, underlying constitutive possibility upon which the expressive forms of feminine masochism and moral masochism are constructed, and Antonio's own masochism arguably partakes of both forms. Fantasies of feminine masochism "place the subject in a characteristically female situation; they signify, that is, being castrated, or copulated with, or giving birth to a baby."[36] Of course for Freud, such gendered categories are by no means tied to the biological sex of the masochist in question; indeed, he commences his analysis of feminine masochism with the blithe assertion "we have sufficient acquaintance with this kind of masochism in men (to whom, owing to the material at my command, I shall restrict my remarks)."[37] Thus Freud's very elaboration of "feminine masochism" as a position within fantasy is underwritten by a clinical observation of these fantasies in men.[38] While we may bracket the assumption that to imagine one's self as castrated is necessarily to imagine one's self in a female position, the specific scenarios elaborated in this account of female masochism are nonetheless potentially useful for further clarifying in what sense Antonio's reaction to his predicament can be termed masochistic. His identification with this "feminine situation" occurs directly within the play. In the process of justifying the swift execution of Shylock's bond, Antonio specifically defines himself as castrated:

> ANTONIO: I am a tainted wether of the flock,
> Meetest for death. The weakest kind of fruit
> Drops earliest to the ground, and so let me. (4.1.114–116)

The shock of this designation enters the play with the intrusive force of fantasy that Torok describes. What could such a claim literally mean, and how might it clarify the connection between melancholy and masochism?

In describing himself as a "tainted wether of the flock," Antonio explicitly takes up the masochistic fantasy position which Freud found definitive of "female masochism": He imagines himself as already castrated and thereby weakened and unfit to live. What is interesting is the circularity with which Antonio asserts this imagined status in order to reify rhetorically Shylock's claim upon the pound of flesh. When naturalized into the logical outcome of his castrated status, the evisceration and removal of the pound of flesh is framed as somehow secondary to an original act of violence projected proleptically forward, transforming Shylock's cruelty into both its double and its justifiable, inevitable consequence. The extraction of the pound of flesh will castrate Antonio for the second time, a fate he deserves because he has already suffered it.

As numerous critics have been quick to point out, Antonio's comparison of his melancholic body to a castrated male sheep recalls Shylock's taunting deployment of the parable of Laban's sheep earlier in the play; but if such a link is easily established, its particular meaning remains elusive. M. M. Mahood's editorial wariness in glossing the line is typical: "Antonio may mean he is expendable because he has no family of his own. But the image may relate to Shylock, seen either as a shepherd culling a useless beast from the Christian flock (suggested by his talk of Jacob in 1.3) or as a dog worrying sheep."[39] This is the problem of the "tainted wether" epithet: It prompts the question of whether the "taint" figures the proleptic cause and justification of the castration, or its consequence. Antonio describes the weakened and disabled melancholic body as somehow not generative, infertile, incapable (and therefore unworthy) of breeding. Such a claim provides a crucial insight into the relationship of mutual support between Antonio's melancholy identification and his embrace of the masochistic fantasy that surrounds him; the two are connected in the flamboyant proclamation of his castrated status, fusing until castration comes to figure both melancholic cause and masochistic outcome. Answering the question "Is there a Cause of the Subject?," Slavoj Žižek figures this logical loop within the causal sequence: "[A] certain radical ambiguity pertains to cause: [C]ause is real, the presupposed reef which resists symbolization and disturbs the course of its automaton, yet cause is simultaneously the retroactive product of its effects."[40] This temporal circularity (I deserve to be castrated because I am already castrated) partakes of the formal endlessness already encountered in the speculative interpretation of melancholy causality: "Are you sad because of X?" "No." "Are you sad because of Y?" "No." "Are you sad because of . . ." and so forth. But it also exhibits the repetition characteristic of masochistic fantasy as it lurches toward, yet prolongs the suspenseful anticipation of, end-pleasure. Masochistic sexuality obtains in the ritualized threat of pain as much as in the experience of it: The "to be" of the castration-to-come must always be imminent but never arrive.

We can now see that the masochistic fantasy of being "already" castrated symptomatically literalizes at the level of the body what was registered in melancholy as an epistemological loss. The feeling that "there is something missing, something that I lack" has been translated from an affective loss (I no longer know why I feel sad, I have lost self-access) to a physical loss (I no longer have a body part, I have lost the phallus). But the transition of Antonio from melancholy to masochism does not simply rearticulate the "same" affect through a change in his behavior, it also introduces a surplus of pleasure or enjoyment in the shift from the epistemological to the somatic. Furthermore, the admission that "I am a tainted wether" constitutes a kind of melancholy self-knowledge which suggests that, through his entrance into the masochistic fantasy with Shylock, Antonio has come to a kind of self-consciousness that revises and replaces the melancholy nonknowledge that commences the play. What was an ignorant blind spot has now been filled in with a bitterly pleasurable "truth": castration.

While Antonio's masochism fits the contours of the Freudian paradigm of "feminine masochism" in its fantasy of castration, it has other features that align themselves more productively with Freud's final account of "moral masochism." In contrast to the erotic and sexualized scenarios of feminine masochism, moral masochism, with its calculus of unconscious guilt producing potentially unlimited punishment, places the specific identity of the agents at an abstract remove and focuses instead upon a nonspecific, endless work of self-punishment typically enabled by a quasi-ethical or religious rationalization of one's suffering as the result of one's sinful nature. Freud suggests that this layer of narrative abstraction allows the agents of cruelty within the moral masochist's fantasy to be refigured into the completely impersonal workings of fate:

> All other masochistic sufferings carry with them the condition that they shall emanate from the loved person and shall be endured at his command. This restriction has been dropped in moral masochism. The suffering itself is what matters; whether it is decreed by someone who is loved or by someone who is indifferent is of no importance. It may even be caused by impersonal powers or by circumstances; the true masochist always turns his cheek whenever he has a chance of receiving a blow.[41]

The naturalized imagery of Shylock's cruelty that Antonio's imagination produces (he figures Shylock as "flood," as "wolf," and as "mountain pines" [4.1.71–75]) demonstrates this principle of abstracted agency at work within moral masochism; precisely at the extraordinary moment when he cries, "Let me have judgment, and the Jew his will!" (4.1.83), Antonio has just imagined himself suffering at the hands of the cosmos itself rather than a single businessman. The Christian turn in Freud's account of moral masochism, his conviction that "turning the other cheek" is an expression of a desire for pain, and a means of securing pleasure-in-pain rather than simply an acceptance of it, is developed further toward the end of his essay and should be kept in mind when reading credulous accounts of Antonio's *amor fati* as a pious Christian act of martyrdom. In a direct parallel with Antonio's headlong pursuit of his own extermination, Freud details the lengths to which the moral masochist is prepared to go to secure a "chastisement from the great parental power of Destiny," yielding a confirmation of the power and authority of his tormentor which is also a tribute of love to the super-egoic source of the suffering: "In order to provoke punishment [. . .] the masochist must do what is inexpedient, must act against his own interests, must ruin the prospects which open out to him in the real world and must, perhaps, destroy his own real existence."[42] Such an analysis not only illuminates Antonio's call "Let me have judgment" but also retroactively allows us to see the same dynamic at work in his initial acceptance of the bond in the first place. The very rashness of his generosity takes on the contours of a moral masochism that seeks the fulfillment of the trial scene as a consummation devoutly to be wished, its necessary consequence and underlying goal. The relentless nature of these self-destructive urges also helps to bring into focus why Freud articulates originary masochism in relation to the workings of the death drive, for what is longed for, ultimately, in the harvesting of the pound of flesh is not just suffering, but death.

Sacrifice, Subjection, and the Melancholy Pound of Flesh

Earlier in the play, Antonio's melancholic "blankness" made his body the site of a certain proliferation of meaning; in the trial scene, the subjection of his body to an impersonal, universal logic of a sacrificial law makes a certain pliancy or flexibility of that torture's meaning possible, which allows for this triple exposure of multiple (Jewish, Christian, masochistic) sacrificial logics. What the play makes available here is precisely not an external, safe position of choice between these logics (considering Portia's triumph and Shylock's forced conversion, how could it?) but rather the direct experience of their contest as the final site of an ongoing process of subjection: the subjection of civil subjects to the laws of the city, the subjection of humanity to divine justice. The pandemic of overdetermination which releases the flood of possible melancholic and masochistic meanings is always shown to be subject to the regulatory forces of law that enable some forms and foreclose or forbid others. There is, then, no "alternative" to subjection which the play imagines, no liberating "outside," and the choice between these competing regimes remains, in each case, a choice of who shall be master.

How can we think through this interpenetration of Christian and Jewish meanings in relation to the empty neutrality of a civil order that would seem to stand outside of either of them? Where does the law stand between Christian mercy and Jewish sacrifice? The Christian reading can be rehearsed as follows: As the person who pays the mortal debt of another, Antonio explicitly takes on the position of the suffering, martyred Christ, and he does so in a manner calculated to display "Jewish" cruelty in the process. The pound of flesh is the payment for the sins of others (doubly profane, Bassanio's worldly debts have been incurred for the sake of erotic conquest). These debts are taken up by an asexual, pure, and guiltless sacrificial victim who is threatened with death by the bond that holds him in thrall to a bloodthirsty Jew (though here Antonio's self-designation as "tainted" catches in the throat). If the fantasy scenario of Antonio's death at the hands of Shylock registers as a shocking eruption of alien values (the dead letter of a "Jewish" law) into a Christian community of love and mutual concern for human life, his embrace of the very terms of this ostensibly "alien" contract is experienced as the necessary act that precisely exemplifies and manifests the spirit of the Christian community. Thus the anti-Semitic projection of the inhuman, ravenous Jew serves as a prop or support to Christian self-understanding by negatively consolidating its members as witnesses to Antonio's proposed sacrifice, a sacrifice which, were it to occur, would mirror and cite the exemplary, prior sacrifice of Christ to his Jewish and Roman executioners. But in describing himself as the "wether of the flock," Antonio also seems to occupy, symbolically, the position of the sacrificial animal within Jewish ritual. Though his crucial addition of the word "tainted" disqualifies him from eligibility for such a position, we can nonetheless detect typological echoes of an older, paradigmatic example of sacrifice in Antonio's urge to sacrifice himself: Abraham's willingness to sacrifice Isaac as a sacred test of his subjection to divine authority, with Antonio playing Isaac to Shylock's Abraham. When read in terms of Abraham's test, Antonio is that which must be consumed or destroyed as a sacrifice to a higher principle which is understood as self-justifying, absolute, and, to its horrified Christian onlookers, "irrational" in its literalism; this principle is allied not with a personified agent such as Christ but with the abstract power and absolute necessity of a universally binding Law.

The predominant opposition between the Christian spirit of mercy and the Jewish letter of the law has, at this point in the history of the play's critical reception, collapsed the play's complexity into an altogether too-pat collection of Pauline theological commonplaces. Perhaps what is needed is a new way to think about the status of subjects as bodies under law, with subjection redefined here as the mark of being subject to an external authority that conditions and makes the subject possible through a network of laws that both protect the body and subject it to universal codes of regulation, punishment, and control. The promised spectacle of Antonio's "fair flesh" being subjected to the law and forced open initiates a crisis for the civilizing regime of Venice, for it promises to rend and reveal the body's carnality, its thing-like priority to any symbolizing and civilizing process, and if Shylock receives his bond, that regime will be made to do so precisely in the service of maintaining civil order. If Shylock does not receive his bond, then the laws of Venice are null; the revelation of the pound of flesh, which the maintenance of civil order requires, will itself reveal the murderous violence inherent in the exercise of civil order, and so discredit the very state it is meant to legitimate.[43] Antonio risks exposing the "Jewish" character of Venetian civil order, that is, the rigorous subjection of its citizens to the letter of the law. In this sense, law is experienced as simultaneously foundational (it provides the rules by which Venetian society is to be ordered) and foundationless (it cannot, itself, be justified; it simply is the law). Thus, the tautological bedrock of the law, its foundational status and universal applicability, is both "full" (it saturates the social field and must apply absolutely and in every case; otherwise it would not be the law) and "empty" (it rests upon nothing but its own self-identity, its own status as itself-as-law, and cannot be propped up by anything outside).

Shylock's urgent claim "I stand here for law" (4.1.142) telegraphs his hopes that he can dissolve the partiality of his suit into an equation with the impersonal rule of law itself. Of course, when read in terms of the standard Pauline oppositions both complicated and kept in play by recent Shakespeare criticism, in speaking as a Jew, Shylock already "stands for law," and in a manner calculated to centrifugally oppose his "Jewish" legalism to Antonio's supposedly "Christian" self-sacrifice.[44] But we may also experience this claim as Shylock striving to dissolve his very Jewishness into the categorical blankness of membership in a generalized citizenry; he is appealing to his right to the bond not as a specific Jewish usurer but as an impersonal, abstract subject of the law. Shylock asserts precisely the universal applicability of law to both citizens and "strangers" in order to gain leverage against his religious and economic enemy, only to discover through Portia's final *coup de grace* that, in fact, that law is riven by distinctions between aliens and citizens, and that these distinctions can and will be used against him.[45] Portia's celebrated excursus upon mercy would seem to suggest that there is a principle prior to law which stands ready to rescue us from the choking grasp of the literal, yet her own merciless application of Venetian law against Shylock places mercy outside the text (it is, precisely, mercy which is "alien").[46] Portia's insistence upon the literal flatness of the law (exactly one pound, no more and no less) is a joke at Shylock's expense which recalls Deleuze's assertion that "irony and humor are the essential forms through which we apprehend the law."[47] But the larger joke in the background is that mercy is only ever the gift of power, and Shylock is not entitled to share in that power. Portia's urgent command to carve up Antonio and her show-stopping reversal of that command come from the same authoritative position. Dr. Balthazar's ostensible grounding in the text of Venetian law (itself an act of imposture that dissolves Portia's responsibility into so much comic drag) just barely conceals the lingering threat that that which has been reversed here could also be reversed again, and so on, endlessly, and yet in either case her decision would be just as demonstrably founded in law.[48] This neutrality demonstrates the powerlessness of power to interpret itself. As a code which is always hypothetically reversible, always to-be-applied, the law generates and makes possible a manifold of different forms of subjection (sacrificial victim, exemplary case, anecdotal precedent, guilty criminal). Subject to interpretation, the law can also be turned and turned again to serve the interests of the very individuals it subjects; it is a template whose blank neutrality enables Antonio's embrace of the masochistic fantasy, Shylock's pursuit of his sadistic lust for revenge, and Portia's bitter reversal of both their desires.

What would it mean to "stand for law"? What sheltering camouflage might the impersonal rule of law provide? Gilles Deleuze has depicted the hyperconformist embrace of the law as a critical act of identification with the law concealed within postures of obedience to it:

> By observing the very letter of the law, we refrain from questioning its ultimate or primary character; we then behave as if the supreme sovereignty of law conferred upon it the enjoyment of all those pleasures that it denies us; hence, by the closest adherence to it, and by zealously embracing it, we may hope to partake of its pleasures.[49]

Deleuze's insight helps us to grasp Portia's adherence to the letter of the law during the trial scene as a parodically pleasurable identification with law's arbitrary sovereignty. Portia's execution of her father's will, with its chastely modest fulfillment seasoned by wisecracking asides, prepares the audience for this troubling identification with an absent yet all-powerful authority by modeling an obedience which is *both* humorously subversive and yet faithful to the very letter. Portia is compelled from without by her father's will, but what the play shows us in the Belmont scenes is that each moment of obedience to that will is itself an active moment of choice. Agency is constituted in and by a subjection that is not merely external and separate from agency but is itself one of its modes of expression. Considered from a position in which compulsion and choice actively interpenetrate each other, the weary critical debate about whether or not Portia fixes the outcome of Bassanio's choice of caskets can be seen to rest upon a more troubling, and more interesting, underlying problem: freedom and its pleasures issue from a constitutive underlying subjection. Read this way, Portia's occasional protestations of revolt against the compulsion of parental authority scan as a technique for muffling or concealing the intolerable pleasure of obedience. This dynamic is not limited to Portia. According to the schematic of the play's "apparent content," Antonio is compelled from without, forced to submit to the threat of a terrifying ordeal. Yet, as with Portia's active obedience, the play also shows the extent to which Antonio has chosen that compulsion, reveling in the impersonal authority of the law while turning it to serve the ends of his anti-Semitic hatreds and masochistic scenarios, using it to ratify his melancholic self-understanding as "meetest for death."

Antonio's cry "Let me have judgment, and the Jew his will" (4.1.83) troubles the surface plot of legal compulsion by revealing Antonio's predicament as an obscenely pleasurable masochistic choice. The law compels a decision, but the vigor with which Antonio calls for his sentence somehow exceeds the law that occasions it. This is why the fantasy of the extraction of the pound of flesh creates such intolerable anxiety within the play, and why the threat of its realization in the courtroom can only be assuaged through the emergency re-inscription of a generalized subjection to the rhetoric of mercy, prompting Portia's desperate yet dazzling campaign to close over the trembling void of the pound of flesh. In striving to have his interiority torn open for inspection, Antonio's masochistic embrace of this fatal scenario represents the apotheosis of the melancholy self-anatomizing that begins the play. What began as a longing to have his suffering understood by his fellow citizens and comforters becomes a fantasy scenario in which those citizens are nearly forced to preside over his suffering and death in a gruesome spectacle that threatens to rupture and discredit the civic bond itself. When we read his masochism and his melancholy together as deliberately incomplete epistemological assemblages whose provocative gaps and troubling spectacles allow the question of desire to be raised, Antonio's curiously exceptional status as a satellite to the marriage plot dissolves. What comes into view instead is a critical sense of Antonio's melancholy and masochism as modes of subjectivity whose insistence upon participation and collaboration force into the open the disavowed conflicts at the heart of the society that surrounds him. Instead of an isolated, inconsolable figure at odds with the heteronormative comedy in which he suffers, we can see the symptomatic logic through which Antonio's showy declaration of his sadness and his frightening embrace of his own domination mirrors the broader forms of subjection that drive the subjects of Belmont and Venice throughout the play.

In reading Antonio's melancholy and his masochism together as an index of a general logic of subjection which sustains familial, economic, political, and religious relational networks, I hope to banish the spectre of interpretive foreclosure posed via Turner at the beginning of this chapter. Far from tunneling inward into an abyssal privacy that ignores economic and political meaning, psychoanalytic readings might provide the means by which to discover the exceptional, sacrificial, and symptomatic logic(s) of fantasy within the political and economic contours of lived experience. In the workings of Shylock's economic contract, in Portia's obligation to her father's will, in Bassanio's pledge of marital fidelity, and in the Duke's need to uphold the laws of Venice, subjection saturates the social assemblages of *The Merchant of Venice*, determining Antonio no more and no less than his fellow citizens. My expansive use of the term *subjection* here rests upon the assumption that to occupy a legible place within the social is to remain answerable to the forms of authority which ground it: "To be a subject" and "to be subject (to authority)" are more than kin and less than kind. In this minimal sense, "subjection" marks the point at which personhood and citizenship border and presuppose each other; the exceptional status of refugees, aliens, and criminals indexes the grey zone of subjective disappearance at the borders of authority. Much of the comic energy and nerve-wracking momentum of *The Merchant of Venice* stems from the ingenious means through which this constitutive sense of subjection (on its own, a merely grounding fact that the self is always a self-in-society) metastasizes as arbitrary, cruel, or questionable demands are placed upon the characters within the play. The will of the father that enforces lifelong celibacy upon those who fail to guess the casket riddle, the terms of the contract that stipulate the pound of flesh, the final sentence of forced conversion—these outcomes render the status of daughter, bondsman, or citizen problematic by pushing the obligations entailed by each term to their limits. Seemingly compelled by forces beyond his control, Antonio's cry of desire—"Let me have judgment, and the Jew his will" (4.1.83)—breaks open the conflicts at the core of *The Merchant of Venice* as a whole, bringing its constitutive fault lines (between Christian and Jew, master and bondsman, letter and spirit, Venice and Belmont, economics and politics) into critical articulation, revealing the assemblage that binds melancholy epistemology and masochistic fantasy together.

4. *That Within Which Passes Show*

Mourning, Melancholy/Melancholia, and "the Hamlet mystery"

In a farcical piece of stage business with a flute, Hamlet issues a famous rebuke to the courtier spies Rosencrantz and Guildenstern that applies equally to the spectators, readers, and critics who attempt to cash out the meaning of his melancholy:

> HAMLET: Why, look you now, how unworthy a thing you make of me! You would play upon me, you would seem to know my stops, you would pluck out the heart of my mystery, you would sound me from the lowest note to [the top of] my compass; and there is much music, excellent voice, in this little organ, yet you cannot make it speak. 'Sblood, do you think I am easier to be play'd upon than a pipe? Call me what instrument you will, though you fret me, [yet] you cannot play upon me. (3.2.343–351)

Hamlet works to create a desire for the articulation of a buried content whose expression he teasingly withholds. There is a temporal structure to this promissory dynamic; revelation is always at hand—but not yet. Interpretation is the occasion for a kind of performative shame at the incapacity of some other (impressed into service as a local representative of a Big Other of "court," "the world," "Denmark") to penetrate the occulted interiority of the melancholic figure—who places himself on display—while also decisively closing himself off from any access. This shame is ostensibly experienced by Rosencrantz and Guildenstern, but the reception of the play has considerably amplified this pervasive, unwelcome emotion. Hamlet understands himself as the site of a particular projection of his own spectator's interpretive desires, and works to rhetorically inflame and provoke those very longings even as he forecloses their possibility and preemptively shames his would-be interpreters; but his aggression in doing so marks his discourse with eruptions of desire and lines of force that may, all the same, be telling in their own symptomatic manner.

Faced with such a deadlock, what is needed may be a certain salutary resistance to interpreting Hamlet's abyssal interiority where it announces itself directly, a certain refusal to take up the problem of his "mystery" as he has conveniently staged it for our consumption. Margreta de Grazia's provocative *Hamlet without Hamlet* has recently attempted to do just that, and yet the subsequent responses to de Grazia's historicist intervention have demonstrated that her attempted shift toward the political dimensions of Hamlet's disinheritance from Denmark's succession constitutes an additional motivational gloss upon, rather than a replacement for, the enduring mystery of Hamlet's inwardness.[1] Laying springes to catch his would-be interpreters, Hamlet has played a part in fashioning the very topology of abyssal subjectivity which conditions many of the critical attempts to pluck "out" the heart of a mystery felt to be somehow "in" an epistemological strongbox bounded by his person. Responding to Gertrude's query about his persistent gloom with an aggressively self-righteous attack upon her, Hamlet forcefully announces an absolute epistemological division between an outside and an inside, and uses this division to subtly undercut her mourning as inadequate when compared with his unverifiable yet insistent proclamation of his own authentic mourning. It is a response pregnant with meaning, yet maddeningly empty:

HAMLET: Seems, madam? nay, it is, I know not "seems."
'Tis not alone my inky cloak, good mother,
Nor customary suits of solemn black,
Nor windy suspiration of forc'd breath,
No, nor the fruitful river in the eye,
Nor the dejected havior of the visage,
Together with all forms, moods, shapes, of grief,
That can denote me truly. These indeed seem,
For they are the actions that a man might play,
But I have that within which passes show,
These but the trappings and the suits of woe. (1.2.76–86)

It is all here. In a productive, indeed foundational, conflation of the two states, Hamlet offers a blazon of mourning which is also a blazon of melancholy, itemizing a full repertoire of external signs and symptoms, from the physiological (tears) to the affective (psychological dejection) to the sartorial (black clothing), which the two conditions share equally. In a play which obsessively restages the iconographic depiction of the melancholy figure, Hamlet's founding act is to self-anatomize his own conformity to a public and discursive framework which allows his mourning *and* his melancholy to be detected and recognized by others, and keeps them in a provocatively indistinguishable proximity. But this self-description, which triggers a paratactical outpouring of dependent clauses that keep limning his portrait with extra details, is also shot through with a tic of negation, a "no" that precedes each symptom as it is checked for, discovered, and then scratched off the list. Hamlet "has" all of these things, but none of them are sufficient to verify or support the mysterious *X* which is their absent cause, which he is also said to "have." In the very teeth of this showy denial that any of these outward signs are trustworthy, Hamlet asserts possession of a thingly "that" which bears no nomination, no articulation, no exposure. Furthermore, the nomination of "that within" in a speech produced here for Gertrude's and Claudius' and our own audition marks not the assertion of a fact but the expression of a wish: This speech does not establish that Hamlet has anything at all, but it clearly telegraphs that Hamlet wants us to think that he has something. Even as it betrays Hamlet's desire, this speech works to create a certain desire on the part of its listeners; Hamlet wants us to parse "the Hamlet mystery" as a direct and simple interrogative: We are meant to ask "what exactly *is* that within which passes show?"

According to one popular projection onto this representational black hole, Hamlet's "that within" denotes a rationally argued commitment to the verifiable existence of a self-transparent, privately accessed subjectivity, and one that occupies a uniquely privileged position in the cultural and intellectual history of the West. As such, it has more recently become the target for a clamorous chorus of philosophically skeptical argument, critical and historical deconstruction, political critique, economic positioning in relation to class, and corrective accounts of its complicity in the tacit masculine gendering of the rational subject.[2] Such responses, when launched in broad opposition to the target of Cartesian subjective rationality embodied in the "I" which apprehends itself in the *Meditations*, are each as individually successful or unsuccessful as their particular imaginative and argumentative strengths afford. But they share a common characteristic: their target is and is not, quite, Hamlet. Hamlet's very refusal to provide a legible explanation of what his "that within" consists of, indeed his aggressive overproduction of possible but optional explanations for that question, when combined with his ducks and feints away from explanatory closure throughout the play, may share a family resemblance with the cunning exit strategies and rhetorical soft spots detected within the Cartesian project. But Hamlet's bare claim "I have that within which passes show" does not necessarily commit him to a psychic forecast of full dress Cartesian rationality any more than it commits him to a nostalgic endorsement of feudal *virtu*. What has been arguably lost in this ambitious assimilation of dramatic character to a ready-to-hand philosophical position is the obstinate emptiness of Hamlet's bare ostensive gesture when measured against Descartes' resounding affirmation.[3] To avoid a retroactive projection of Cartesian subjective rationality onto the humoral subject incarcerated in the prison of Denmark, I intend to recover the ostensive force of Hamlet's "that" and foreground its thingly, material character by making an ambitious assertion of my own: The "that within which passes show" is melancholy.

Lest this seem like merely another entry in the critical guessing game that Hamlet's bait-and-switch occasions, or, worse, an obstinate failure to see that Hamlet is simply asserting that he feels genuine grief at his father's death and no more than that, consider the immediate dramatic context which has produced this extraordinary announcement that he possesses that within which passes show. According to Claudius and Gertrude (who is gentler in her critique but equally threatened by the implied censure of his behavior), Hamlet's mourning has surpassed its occasion; it shows up as excessive and "unmanly." Insofar as it now warrants explanation and excuse, "mourning" no longer seems adequate; indeed, the very question "*why* do you continue to mourn?" implies that there is, in fact, another emotion which lies concealed beneath the occasion that mourning here provides. "I mourn because I mourn" would be tautological, and no answer at all. Out of mourning, Hamlet has made a nest in which to nurse a something else, an affective remainder that requires explanation.

Freud's essay "Mourning and Melancholia" takes up at its starting point the curious similarity between these two states, and usefully articulates both the structural dependence of melancholia upon mourning and the excessive, seemingly unmotivated nature of the loss registered by melancholia in a manner which is distinctly resonant with Claudius' and Gertrude's enquiries and Hamlet's response to them. Mourning and melancholia are said to resemble each other, or, more accurately, to "seem to correlate," "whenever it is possible to discern the external influences in life which have brought each of them about."[4] At this point in the essay, it is mourning which functions as the pattern upon which melancholia takes its imitative form, for when he describes what that external influence consists of, Freud shifts from a comparison between the two terms to the single elaboration of mourning: "the reaction to the loss of a loved person, or to the loss of some abstraction which has taken the place of one, such as fatherland, liberty, an ideal, and so on."[5] Here already we face the possibility of a kind of virtual substitution of persons by ideals within the original definition of mourning itself. Before melancholia has entered the picture then, mourning itself already takes on an ambiguous status with relation to its object: What is original or primary may already itself consist in the mourning of a substitution or placeholder occupying the position formerly taken up by a loved person. Described as an "effect" of the same influence of loss, in some persons melancholia occurs "*instead of* a state of grief."[6] Thus, for a select group of sufferers, there is a similar kind of substitution occurring precisely in their entrance into melancholia rather than mourning; melancholia is taking the place reserved for mourning, it is occupying its position. Given the particularly vexed circumstances of the "falling off" occasioned by "his father's death and our o'erhasty marriage," it is tempting to diagnose that Hamlet's reaction has been precisely such an entrance into melancholia rather than normative mourning, and the first wave of psychoanalytic literary critics of the play (Ernest Jones, Ellie Sharpe) have more or less done so.[7] Freud then shifts his ground, remarking that grief in reaction to loss is an entirely normal, expected, possibly required reaction, and that no attempt to intervene or curtail this normative grief should be made. The work of mourning is the painful process of removing one's libidinal attachment from the lost object, a slow and costly struggle in which each memory of the lost person is brought forward and harrowed and, in terms of libido, annulled or cancelled as a site of investment for "the economics of the mind"; the mourner "pays" respect to the lost person in order to void their account: "Each single one of the memories and hopes which bound the libido to the object is brought up and hyper-cathected, and the detachment of the libido from it is accomplished."[8] In asserting that by the end of a month Hamlet should have fully accomplished this process of detachment and acceptance, Claudius and Gertrude are enforcing a particularly hasty rhythm onto the process of decathexis, a tellingly forced expediency. When measured against the timetable of their own desires to be rid of the memory of Old Hamlet, Hamlet's persistence in grief shows up for them as a provocation and a mystery.

It is precisely in such terms that Freud articulates the distinction between mourning and melancholia: Their difference is not a difference in the nature of the loss but in the nature of the consciousness of the loss, and melancholia is defined by a kind of mystery, an interpretive task posed by the simultaneous awareness that a loss has occurred and confusion as to precisely *what* has been lost: "The object has not perhaps actually died, but has become lost as an object of love."[9] Furthermore, this mystery may obtain for the psychoanalyst, or the patient, or both. Patterned upon the brutal simplicity of grief, melancholia borrows its affective repertoire, but its apparent "baselessness" presents itself to the melancholic subject and those around that subject as a cause for interpretation, a source of doubt. As Freud describes this difficulty: "[T]he inhibition of the melancholiac seems puzzling to us because we cannot see what it is that absorbs him so entirely."[10] Here the resonance between Freud's account of the puzzled analyst struggling to understand the nature of the melancholiac's loss and Claudius and Gertrude's struggles to understand the persistence of Hamlet's grief closely overlap—*even though the initial nature of his loss seems clear.* There is no doubt that the "lost object" is Old Hamlet, at least according to Hamlet's initial soliloquies and speeches (though the idealization of his father into a Hyperion also indicates that a certain "abstraction" of the dead loved one into an ideal rather than a person is already at work here, and has yet to undergo the crisis of the Ghost's revelations).[11] But Hamlet's reaction puzzles those who observe it, for it seems to go beyond the occasion of mourning, and it routes their (and our) interpretive energies as an assemblage of spectators in ways that already work to produce or trigger the ascription of a buried "that within," before Hamlet's show-stopping announcement enforces that topology publicly.

Braving the fatal effects which a failed attempt to pluck out "the heart of the mystery" occasion within the play, there has been no shortage of readers and critics who have picked up the task which Rosencrantz and Guildenstern have left unfinished: The quest to score Hamlet's "most excellent music" and draw out his mystery has produced a floodtide of Hamlet (and *Hamlet*) interpretation and analysis, a kind of Niagara Falls of critical discourse plunging into the void which Hamlet's gesture toward a buried affect keeps open for inspection. Given the extraordinary breadth, diversity, and longevity of the secondary and critical literature that comes in the wake of the play, any survey of its critics and readers must necessarily be strategic and selective. To risk stating the obvious, one potentially useful line of distinction that the problem of "the Hamlet mystery" radiates across this critical panorama is the bright line which separates those who believe *that there is* a "Hamlet mystery" from those who do not, such as de Grazia. It is fair to say that much of the criticism that this play has generated has been taken up with fashioning a response adequate to the interpretative provocations staged by this character, and that many of these responses have been proposals for an at least partial resolution to that "mystery" or provocation of some sort, however shot through with anxious throat clearing and hedged round by critical gestures of modesty. The critical continuum that extends from Goethe, Coleridge, and Hazlitt up to Bradley and on into the "first wave" of psychoanalytic criticism (Jones, Sharpe) shares in this positivist spirit: As wildly divergent as they are from one another, they are all direct essays into the rough and murky terrain of Hamlet's psyche or personality in search of a core characteristic, an adequate "thick description," a clarifying structural answer, a binding together of operative tensions, a satisfying account of at least the predominant forces at work "within" Hamlet's interiority. In short, they take up the opposition between what is "outside" and what is "inside," which Hamlet makes available as his "mystery," and work their way "in" accordingly.

But if the extraction and advancement of a particular doctrinal or philosophical position from amidst the scattered, manic soliloquizing and endlessly quotable *bon mots* of Hamlet's speeches within the play constitutes the bravely positivist response to "the Hamlet mystery," an equally popular route has been to advance a dogmatically skeptical refutation of the view that there *is* an available solution. T. S. Eliot, John Dover Wilson, and Francis Barker have all advanced notably agnostic responses to "the Hamlet mystery," and before further advancing my own reading of the experience of melancholy within the play, I shall here attempt, as briefly as possible, to rehearse the variations in their modes of negation. The movement across these three responses demonstrates a certain sequence of maneuvers and shared assumptions which prepare the way for my own assessment of the expanding ambition and increasingly disturbing experience of Shakespeare's use of melancholy in *Hamlet* as he subjects Hamlet's claim to possess "that within which passes show" to dramatic scrutiny.

Now a Mona Lisa of criticism to match its own famously feline assessment of Shakespeare's play, T. S. Eliot's essay "Hamlet and His Problems" (1920) seeks both to devalue the play (*Hamlet* is unsuccessful as a work of art) and to foreclose the possibility of understanding it (*Hamlet* is based upon an inexpressible emotion). The two claims are related: It fails as art *because* it turns on an emotional reaction to an "experience" that cannot be known. The core problem of the play is thus an epistemological gap in understanding the basis of an affective display, and one that promiscuously doubles itself, with our own impossible understanding of the play stemming from a prior incomprehension of Shakespeare's. According to Eliot, to understand the play, rather than to dangerously admire one's self in it (à la Coleridge or Goethe) or to find it merely and vulgarly "interesting" as the masses do, we would have to supply a content to fill a lack within the text, and in order to do this, "we should have to understand things which Shakespeare did not understand himself."[12] Eliot projects a lack of self-knowledge onto Shakespeare, and then makes Shakespeare responsible for the consequent reader or playgoer's corresponding lack of knowledge. Moving in for the kill, Eliot acts both pained and relieved at this lacuna: "We must simply admit that here Shakespeare tackled a problem which proved too much for him. Why he attempted it at all is an insoluble puzzle; under compulsion of what experience he attempted to express the inexpressibly horrible, we cannot ever know."[13] Eliot does not qualify his assertion with reference to cultural or historical or clinical barriers to knowledge—the bar is absolute. While he grants that there are gaps in the textual record of Shakespeare's life, the problem of knowledge occasioned by *Hamlet* transcends the reach of any possible biographical discovery. Eliot forecloses the possibility of such knowledge by attributing an impossible character to the object of knowledge which Hamlet both solicits and conceals: "We should have, finally, to know something which is by hypothesis unknowable, for we assume it to be an experience which, in the manner indicated, exceeded the facts."[14]

What is most unusual about Eliot's reaction is that it seems to describe a rather successful evocation of melancholy on Shakespeare's part. As articulated within his own essay, Shakespeare's capacity to produce the very pervasive sense of an intensely felt but unspeakable affective charge, a feeling that solicits our interpretation while failing to ever quite make itself directly manifest, sounds like a successfully articulated "objective correlative" for the evasive retreat, strangled half-confessions, and unspeakable inner malaise which constitute the experience of melancholy. How has Shakespeare "failed," according to Eliot? The answer, in a word, is timing. The temporality of the objective correlative is for Eliot necessarily instantaneous, an intrinsic and immediate correspondence of emotion to form or object. *Hamlet* supposedly violates this by not communicating directly through a successfully signifying pattern or "chain of events" or object, but through the nebulous miasmic cloud of an underlying tone, a pervasive sense or feeling of intensity which is marked as inadequate to its dramatic occasion. Commencing the skeptical critical tradition with regard to "the Hamlet mystery," Eliot asserts that a constitutive epistemological impasse within Shakespeare's own mind has led to a corresponding representational gap or void in the text—but in the process of making this case, Eliot's essay betrays rather the success of the play at creating for its readers and audience members a dramatic experience of melancholy as precisely that "mystery." He seeks to dissolve the mystery but in fact simply describes it.

Coming from an entirely different position in his evaluation of the success of the play as a work of art, in his *What Happens in Hamlet* (1935), John Dover Wilson nonetheless weirdly echoes Eliot's negative assessment of the Hamlet mystery:

> . . . we were never intended to reach the heart of the mystery. That it has a heart is an illusion; the mystery itself is an illusion; Hamlet is an illusion. The secret that lies behind it all is not Hamlet's but Shakespeare's: the technical devices he employed to create this supreme illusion of a great and mysterious character, who is at once mad and the sanest of geniuses, at once a procrastinator and a vigorous man of action, at once a miserable failure and the most adorable of heroes. The character of Hamlet, like the appearance of his successive impersonators on the stage, is a matter of "makeup."[15]

Coming over two hundred pages into his *explication du texte*, Wilson keeps this deflationary moment well hidden. Hamlet is a matter of makeup and nothing more, and there is no mystery, only a few stage tricks. What more can be said? Such a suicidal disavowal of the interpretive energies of criticism itself must be read as a symptomatic performance of the transmission of melancholy from the stage melancholic to his spectators; what is infectious, above all, is the hollowed-out sensation of a failure in meaning, a bald assertion redoubled by its autumnal tone into something that seems to protest too much. Far from a Prospero-like "drowning of the book" for reasons of ethical restraint ("I could tell you such things but I am sworn to silence"), Wilson's own critico-melancholic performance here indexes a certain tellingly symptomatic fidelity to the object, a Horatio-like willingness to drink of the "same" cup of negation which has felled his idol. As such, this moment recalls Eliot's essay in its symptomatic reinforcement of mystery at the very moment at which it seems to disavow its effect. Just as Eliot's essay claimed that the emotion at the core of Hamlet remained stubbornly "inexpressible" and then proceeded to describe the transmission of that very emotion, so too Wilson's repudiation of "mystery" here subtly works to preserve and extend that very quality, insofar as the assertion that Hamlet binds together contradictory qualities does not provide a resolution to the questions of what those qualities are (those provided by Wilson are epithets of praise and blame), or of how Shakespeare succeeds in balancing them.

This anatomizing of a character into its constituent technical devices is a refusal of "depth" in response to the temptations to read historical context and psychological structure into Shakespeare's creation. Wilson rejects any reduction of Shakespeare's achievement in *Hamlet* to a matter of traditional sources or structural dependence upon an underlying psychological schematic:

> [. . .] though Shakespeare relates his Hamlet to contemporary notions about melancholy, just as he sets his Ghost in a framework of references and allusions to the demonology of his day, neither is composed according to any prescribed pattern or recipe; and they are as greatly mistaken who seek the origin of Hamlet's character in Elizabethan psychology as those who attempt to fathom it in terms of Freudian psychopathology.[16]

In order to secure a Hamlet "flat" enough to be reduced to makeup, Wilson is keen to remove Shakespeare's play from the temporal trajectory of intellectual history, and so he boldly attempts to rescue Hamlet both from the prevailing psychological models of Shakespeare's time and from the terminal moment of his own period's collapse into the profane illumination of Freudian thought. Being "for all time" in the eternal stasis of the self-sustaining aesthetic immanence of the New Critical literary object would presumably rescue the text from the ravages of such reductive understanding. *Hamlet* is not to be translated into other registers but to be structurally described in terms of its productive tensions. Sharing a family resemblance with Eliot's earlier appraisal despite their strong differences, Wilson's relocation of the "secret" from a secret of Hamlet's to a secret of Shakespeare's repeats Eliot's relocation of the central mystery from a gap in Hamlet's knowledge of himself to a founding gap in Shakespeare's knowledge of himself. Furthermore, the two both seem to grant the existence of a mystery even as they struggle to negate it. Wilson's reduction of Hamlet to a matter of balance between incommensurably opposed traits and qualities proposes a de facto technical resolution: We can track the devices, list the traits and qualities, and the structure will emerge. But this comes hand-in-hand with the assertion that Shakespeare remains in possession of a secret, namely how to achieve this delicate balance.

One would expect that fifty years worth of revolutionary transformations in critical methods, fashions, and self-understandings would have produced a vastly different critical landscape. And yet the Eliot and Wilson agnostic response to the Hamlet mystery proved surprisingly hardy. In *The Tremulous Private Body* (1984), Francis Barker argues for a historical shift between two broad epistemes of embodiment which he figures in the specular regime of the Jacobean stage and the successfully privatized, self-regulating, discursively productive Pepysian subject, respectively. Between these two opposed modes of embodiment as subjectively determined by discourse, he locates the figure of Hamlet as the switch point. Hamlet is held up as the figure of incipient bourgeois subjectivity par excellence, a kind of anxious, self-betraying bourgeois subject *avant la lettre*.[17] Curiously, despite his Foucauldian heavy weather, Barker winds up rearticulating the essential claim of John Dover Wilson in *What Happens In Hamlet*—made some fifty years earlier, and on entirely different critical lines—by dogmatically asserting a skeptical nothingness at the core of Hamlet's subjectivity, and reducing it to a promissory bait-and-switch in which "the heart of the mystery" is endlessly promised and endlessly deferred.

Comparing Hamlet with the recorder that Rosencrantz and Guildenstern balk at playing, Barker seems to be ventriloquizing Wilson:

> The hollow pipe is the refutation of the metaphysic of soul which the play signals but cannot realize. For Hamlet, in a sense doubtless unknown to him, is truly this hollow reed which will 'discourse most excellent music' but is none the less vacuous for that. At the centre of Hamlet, in the interior of his mystery, there is, in short, nothing. The promised essence remains beyond the scope of the text's signification; or rather, signals the limits of the signification of this world by marking out the site of an absence it cannot fill. It gestures towards a place for subjectivity, but both are anachronistic and belong to a historical order whose outline has so far only been sketched out.[18]

What "technique" was for Wilson, "history" is for Barker—a solid foundation against which the voids within Shakespeare's text are to be sounded. But Barker fuses Wilson's flattening impulses with a return to Eliot's epistemological void, for according to Barker this heart of nothingness at the core of the play remains steadfastly unknown to Hamlet himself. What is unclear is whether the space of subjectivity to which Hamlet gestures is categorically, necessarily empty, or only empty as a contingent result of his historical displacement from the bourgeois era, which it is his lonely task to prefigure.

Eliot, Wilson, and Barker occupy three distinct modes of skepticism with regard to "the Hamlet mystery," but all three advance the same prognosis for the fate of critical attempts to engage with it. For Eliot, the Hamlet mystery cannot be solved because it rests upon an absolute psychological impasse that has produced a representational failure. For Wilson, the Hamlet mystery cannot be solved because there is no mystery and hence no solution, only an illusion based upon a dramatic technique of balanced yet incompatible qualities. For Barker, the Hamlet mystery cannot be solved because the place of subjectivity to which the play relentlessly gestures is, for historical reasons, empty. With *Hamlet*, Shakespeare perfected a melancholy poetics of generative indeterminacy whose capacity to en-thrall and seduce audience members, readers, and critics is attested to by this character's enduring, vampiric persistence at the center of ongoing cultural and theoretical debates about subjectivity, language, and the nature of the self in relation to family and society. I have gathered together this troika of naysayers because I find their hostility to the entrancing power and seduction of Hamlet's mirage of melancholic depth inspiring and provocative, and I wish to build upon some insights that I find portable in all three of their responses:

1. Eliot's judgment that the play functions largely through a kind of miasmic emotional "tone" that pervades the play but never achieves flashpoint in a particular objective correlative must be both granted and transvalued. Far from constituting a representational failure, this primacy of tone and persistence of diffuse, mysterious, and unresolved affect is in fact a primary virtue of this work, and such a diffusion is in fact characteristic of the early modern understanding of melancholy as an assemblage of persons, languages, objects, and in the dynamic interpretative relations between persons or between persons and objects. In short, if melancholy fails to inhere stably in a precisely articulated "objective correlative," this is not because Shakespeare failed to imagine one, but because melancholy itself is precisely characterized by a *hic et ubique* placelessness and an accompanying feeling of an emotion that strives for release but only seems redoubled by its relentless expression ("as if appetite increased with what it fed on"). Melancholy was understood to circulate in and through elements (earth), seasons (winter), substances (black bile), temperaments (the melancholic personality), moods (the actual sensation of feeling melancholy), and a wide constellation of signifiers, images, and relationships and, in particular, in the cultivation of a theatrical stance that displayed the melancholic individual to society. Thus, Eliot is correct that melancholy lacks a single stable objective correlative but wrong in asserting that Shakespeare's refusal to provide one constitutes an artistic failure. Rather, Shakespeare demonstrates his dramatic understanding of the open possibilities that melancholy's promiscuous significations make possible through his ability to constantly re-present melancholy assemblages across so many diverse strata: objects (skulls), weather conditions (shadowy darkness), emotions (in both manic and withdrawn states), persons (the iconographic/pictorialized images of Hamlet suffering from melancholy as relayed by Ophelia and others), and relationships (the diagnostic frame of interpreta-

tion into which both the court and the audience are placed by Hamlet's provocations). The persistent ambiguity of the very phrase "the Hamlet mystery"—which could designate a mystery within Hamlet the character or *Hamlet* the play—is telegraphed in this case rather neatly; Eliot's objections to collapsing the play into the character are themselves undone by his own essay's symptomatic title ("Hamlet and *His* Problems"; italics mine). This sense of a mystery whose designation endlessly slides between two available locations (an isolated self, or the total structure of the social surround in which that self stands suspended) keeps in play an understanding of melancholy as both a problem to be solved and as a relational network rather than an isolated condition. In this encounter, Shakespeare does not just symptomatically manifest the conflicted intellectual history he has inherited but builds upon it in new ways.

2. Wilson's insight that there are particular "technical devices" within the play which work to construct the interpellating effect of Hamlet's character as it generates a critical desire for understanding and access to a mysterious, occulted interiority is something that I wish to pursue—but not, as Wilson does, in the direction of an evaluative list of the contradictory personal qualities which adhere to his character, but rather through a reading of the dramatic phenomenology of our experience of sharing certain perceptions and intimacies with Hamlet. These experiences occur throughout the play: moments in which we overhear asides which seem to trouble the division between private thought and public speech, moments in which we see and hear something that Hamlet sees and hears but which others seemingly lack access to. These moments constitute a particular relationship between Hamlet and the audience, interpellating us into a kind of ad hoc perceptual community with him—a dramatic assemblage which is based not on sympathy or emotional identification but which seem to occur at some simple perceptual level that precedes consent and resists analysis. In a revision of the standard topology of psychological interiority and discursive/socially legible exteriority, Shakespeare has fashioned a play in which the "inside" view provided by the soliloquies is compounded by an "outside" of diagnostic scenes in which the signs and symptoms of Hamlet's behavior are interpreted by others; then both of these perspectives are further challenged by the experience of an asymptotic approach to the melancholic who is seen and heard "aside," in a kind of sidelong participation in certain moments of shared perception and

audition. As the play complicates the stakes of what is or is not "melancholic" within Hamlet's thinking, speaking, and acting, we are forced to continually reevaluate these experiences of intimacy and community. I take "melancholy" to designate this three-dimensional, dramatic manifold of the "inside," the "outside," and the "aside," and to name the troubling admixture of spectatorship and shared experience which this play makes available for us. It is this particular relationship which constitutes the "generative indeterminacy" at work in Shakespeare's exploitation of certain dramatic possibilities of selective access as they resonate with a constitutive overproduction of possible meanings and significations for this condition/identity in the historical archive of melancholy as a given prop or piece of stagecraft furnished by the contemporary cultural/intellectual history of Elizabethan England. Wilson alleges that the attempt to provide a particular, specific content to resolve the "heart of the mystery" within Hamlet's character is doomed to failure because there is "nothing" there. This bears a family resemblance to my own sense of melancholic subjectivity as a temporal ploy in which content is announced only to be withheld until an endlessly deferred moment of anterior revelation. But this sensation that there is "nothing" there is in fact a consequence of surplus, not lack: there is so much going on that no one "thing," no one single, isolated resolution, is possible. The Hamlet assemblage makes too much meaning available rather than too little.

3. Barker's diagnosis of an "emptiness" within Hamlet that must be understood in historical terms seems to me to be worth preserving, but only if we rethink those terms. In particular, the assertion that Hamlet's "emptiness" is an index of how far ahead of his own cultural moment his incipiently bourgeois subjectivity was must be challenged (if we recognize ourselves in Hamlet, this may say more about ourselves and our understanding of subjectivity than about Hamlet). From Giorgio Agamben's account of the psychological complexity of the medieval monastic concept of *acedia* in *Stanzas* to Margreta de Grazia's wholesale assault upon retroactive critical projections onto Hamlet in *Hamlet Without Hamlet*, recent scholarship suggests that what is needed is a reemphasis on the links connecting Hamlet to the cumulative depth of the melancholic archive that preceded him, rather than a hasty assertion that Hamlet heralds the absolutely new emergence of "our" modernity.[19] While it is incontrovertible that modernity has mobilized Hamlet in its service, and that this mobilization has made

possible the readerly experience of narcissistically discovering our own face within Hamlet's mirror, the availability of this reading must also be rendered rigorously historical, and that cause is not served through loosely typological equations of Hamlet with a particular emergent class.[20] What is needed in order to make good on the linkage between our own modern understandings of subjectivity and Shakespeare's play is a supplementary critical vocabulary: But the vocabulary with which to understand Hamlet's curious persistence and exemplary status as the emergent subject of modernity comes not from Marx but from Freud. *Pace* Barker and, for that matter, de Grazia, it is through a history of the way that psychoanalysis has used and revised its understanding of Shakespeare's play that we can begin to think historically about Hamlet's status as an exemplary model of the emergent modern subject. Through his progressive development of a methodology for reading the logic of the symptom at work within the language and behavior of the subject—and his account of the repressions and resistances and desires at work there—Freud has provided us with the most theoretically ambitious model for how to read "aside" the subject in the manner that I am arguing *Hamlet* itself initiates.

"Less than kin and more than kind": Hamlet's Melancholic Aside

Hamlet makes us his accomplices, joining us together into a socially extended assemblage as he stages a melancholy which we not only witness but cocreate. But how is this possible? Where does it occur? I wish to suggest that the dynamic of the "aside" at work in the play commences when Hamlet first opens his mouth. But this is easier to see in some *Hamlet* texts than in others. Turning to the text of Hamlet's first speech in the play as edited by Susanne Wofford for the "Case Studies in Contemporary Criticism" series, we read the following:

> *KING:* But now, my cousin Hamlet, and my son–
> *HAMLET:* [Aside] A little more than kin, and less than kind.
> *KING:* How is it that the clouds still hang on you?[21]

Turning to the same lines from act I, scene two in the Oxford Shakespeare as edited by Stanley Wells and Gary Taylor, one notes a curious difference, so slight as to be easily ignored:

> *KING:* But now, my cousin Hamlet, and my son–
> *HAMLET:* A little more than kin, and less than kind.
> *KING:* How is it that the clouds still hang on you?[22]

The texts are identical: word-for-word, down to the last punctuation mark, they trace the same sequence of oily courtesy, followed by its coldly witty repudiation, followed by a direct and probing question. But they are also, of course, absolutely different. The presence or absence of the stage direction "[Aside]" marks a radical shift in how we understand Hamlet's first utterance and, I hope to show, opens onto a broader problem of how we are to assess and critically interpret Hamlet's melancholy. Is this first line an aside? If it is an aside, what does that suggest about Hamlet's character, his place within the court of Elsinore, and the dramatic pattern of the play? If it is not an aside, what, in turn, does that communicate? Consulting other editions of the play, one finds a pervasive, and more or less balanced, split on the question of how to identify this first line. Wofford's decision to include the "[Aside]" stage direction places her in the august company of John Dover Wilson, Edward Capell, and Lewis Theobald, while Stanley Wells and Gary Taylor's decision to remove the stage direction in the Oxford edition places them in the equally distinguished company of Harold Jenkins (in the Arden edition) and, more recently, Stephen Greenblatt (in the Norton edition). With an abundance of critical firepower and editorial momentum balanced on either side, the stage is set for a wearyingly inconclusive contention between rival scholarly dynasties.

Rather than contribute yet another polemic in favor of either position as it is currently articulated, I want to suggest that the multiplicity of arguments in favor of both positions about Hamlet's first line should be taken as a symptomatic expression of an underlying problem about the dramatic experience of melancholy. If the inconclusive state of *Hamlet*'s multiple texts prepares the way for critical conflict, the mysterious trajectory of the melancholic address, its capacity to be both for an implied other and yet somehow marked as stifled and private, works in tandem with that larger textual mystery to generate this local impasse. Hamlet's first line shows up as a problem for us as spectators precisely because it seems to hover unstably between private and public utterance; it asks to be located on either side of the very divide which it constitutively blurs. I wish to argue that this speech is in fact an aside, in that it is a speech marked as "private," but that it is specifically a *melancholic* aside, and that as such it must be necessarily public, overheard, and shareable, precisely because melancholy as such abides in an interstitial diagnostic and discursive space between the private self and the social body. Hamlet's line, with its curious toggling between private and public address, is a model melancholic assemblage, for it showily announces the very antisocial withdrawal that it also violates, transmitting inner affect outward, drawing eavesdropping auditors inward. Flagging its aggressively exclusive character in the form of a riddle which is also an insult, his line commences the radically unsettling epistemological abysses that Hamlet's melancholy occasions both for himself and his interpreters within and without the play. The question of what it means for a melancholic to speak an aside is, then, not merely a question of the presence or absence of a particular flyspeck of editorial intrusion upon the Shakespearean page, but another way of posing a more ambitious question: How does melancholy speak? That is, can melancholy be located within discourse and, if so, how? In order to begin to answer these questions, we need to understand the specific contours of Shakespeare's "technical devices" in creating *Hamlet* against the broader backdrop of earlier dramatic traditions in which asides and soliloquies function together to depict subjects as participants in and spectators upon a social realm.

Before determining whether a particular line belongs or does not belong to the category of the aside, the term itself requires clarification. In their coauthored *A Dictionary of Stage Directions in English Drama: 1580–1642*, Alan Dessen and Leslie Tomson have compiled a textual database of hundreds of early modern plays in English, excluding masques and closet dramas, and have constructed a kind of statistical profile of the meaning of particular stage directions. Dessen and Thomson note a pervasive split between an "aside" as a kind of movement and an "aside" as a kind of speech:

> The more than 550 examples fall into two groups: (1) *to stand* or move *aside* so as to observe other figures or carry on some surreptitious activity, (2) *to speak aside* so as to maintain the fiction that the speaker's words can be heard only by the playgoer or by some but not all of the other figures onstage (aside as a noun is not found); [. . .] the majority of asides

> signaled in today's editions are not marked as such in the original manuscripts and printed editions, particularly in the Shakespeare canon where the only designated spoken asides are Quarto *Merry Wives*, E1r, 3.3.139 and *Pericles*, D4v, 2.5.74,78.[23]

As Dessen and Thomson exhaustively demonstrate, with the notable exception of Shakespeare, the deliberate marking of asides in printed editions of plays from the period was quite widespread; G. K. Hunter suggests that the practice of including it in such editions derives from the prior example of classical literature. In his "Notes on Asides in Elizabethan Drama," Hunter collects a wide array of examples of asides from Tourneur, Marlowe, Marston, Heywood, Chapman, Webster, and others, stating that "these not only indicate Elizabethan awareness of the practice in the theatre but also show the range of vocabulary that could be used to describe it," which comprises both inwardly directed asides ("He speakes to himself," "Private," "Apart to Himself") and outwardly directed ("To the People," with "People" designating the audience).[24] Hunter also notes that editions of Terence designed for use in Elizabethan grammar schools included "*ad Spectatores*" to indicate asides, concluding that "any person who had passed through grammar school was likely to have an acquaintance with the aside even if he had been given no generally agreed way of describing it."[25]

If the nominal definition of "the aside" is necessarily incomplete, both because many examples which we think of as asides are not technically designated as such (e.g., the stage direction to speak a line "privately") and because the word itself is used to describe things which we do not really think of as asides at all (e.g., turning aside to make way for another character's entrance), Hunter's assertion stands in need of the supplement of some kind of positive phenomenological description. What is an aside? An aside is an act of concealed speech, and as such it is both intimate and exclusive. At the level of explicit address, some asides are self-referentially addressed to the speaker, some are addressed to another person who is being brought into the speaker's confidence, and some are addressed to the audience. Implicitly, of course, all asides, simply as lines in the script that are spoken with enough volume to be heard, are also always meant for the dramatic audience in at least a weak sense. In order for a line to function as an aside, the speaker must not only successfully reach the audience but the line itself must be seen to fail to register on the part of some other person on stage, who is generally either the gull of an ongoing scheme, the butt of a hidden joke, the object of the speaker's concealed threat, or is him or herself the source of a potential threat were the hidden utterance to be heard. In a sense, this is the aside's unique position at the intersection of epistemology and phenomenology: What we see and hear becomes simultaneously the occasion for a recognition that others have not seen and have not heard, and that, consequently, we now are in possession of a private knowledge which others lack. Merleau-Ponty's observation in *Phenomenology of Perception* that "the act by which I lend myself to the spectacle must be recognized as irreducible to anything else" is apposite here, for it indexes the experiential specificity of precisely such a relation between the observer and the thing observed.[26]

That said, the aside yields itself to further analytic subdivision. Both Dessen and Thomson and G. K. Hunter offer up comparative taxonomies of the aside that draw distinctions between asides which establish a conspiratorial intimacy between characters on the stage and the truly "private" asides in which characters seem to speak only to themselves. This difference in address produces a curious distinction in how we experience the temporality of dramatic utterance. Truly "private" asides slip the timing: they jut out from the normative call and response of dramatic dialogue, breaking the rhythm of its conversational volley. Even—and especially—when they are witty ripostes to what has just been said, they further the course of a play's meaning while violating the implied world of shared discourse in which characters convincingly interact with one another within the play. So asides drive a wedge between the play as verbal pattern or sequence and the play as "world"; they place a stake upon the former at the expense of the latter. The normative conceit of dramatic dialogue is that people speak to and hear one another in a linear, one-way forward temporal flow. But the aside violates this; it is a response but not a response that is delivered back to the person who triggered it and for whom it is (partially) intended. We can assess its meaning, its reasons for not being uttered directly back to its target, and the possibility that it is a borderline aside, mouthed and perhaps just barely overheard by others within the play—only because we as audience members have been included in its radius. This effect of solicitation generates pleasurable and disturbing complicity, but it does so at the expense of the social, which is now recognized as partial, fragile, and subject to manipulation.

When attempting to define the particular effects of the "private" aside, it is tempting to simply annex this category of self-addressed speeches to the more robust and privileged category of the soliloquy, and regard private asides as simply short soliloquies made in the midst of the social surround but to the same purpose as their loftier relative. But the aside is arguably more formally radical in its implications than the soliloquy, more meta-theatrical and, correspondingly, unstable in its effects. By definition, soliloquies occur when other characters are offstage; "*Exeunt omnes, manet Hamlet*" leaves Hamlet alone onstage with his thoughts. Throughout this transition into and out of the soliloquy, there is an implied temporal continuity which remains in effect; the others outside the range of the soliloquy are simply gone, but through a dramatic convention of "object permanence" they are presumed to simply carry on their lives offstage. Hamlet and Horatio's audition of the trumpet flourish and gun salute that announces Claudius' toast in act one, scene four functions in precisely this way, manifesting a perceptual solidarity that links onstage and offstage action together (1.4.7–11). The audience's security that dramatic characters share perceptions which they also experience works to make divergences such as dramatic asides and ghost visions more uncanny, disturbing, and exceptional. When we know that there are onstage listeners with eavesdropping access to a seemingly private speech, the overheard speech winds up affirming the ease of intersubjective access upholding the binding fabric of the social. Both true soliloquies and overheard soliloquies affirm the sense that all of the characters of the play abide in a shared, linear temporal flow. "Private" asides do not do this and can be staged in ways that upset our confidence in the idea of a shared temporality because they raise an awkward question: What are the other characters doing during the time of the "private" aside? If they are miming action and activity more or less silently, their motions tend to convey a kind of implicit critique of the social as surface, as empty speech, and commence a paranoiac-critical stance toward society which is refigured into a kind of dumb show through such a differential "depth of field" (as in the Hamlet-aping malcontent Vindice's opening address to the audience in *The Revenger's Tragedy* [1607]). If those out of earshot are motionless and still during the time of the aside (the "freeze tag" approach to drama), this has the effect of interiorizing the aside itself, of pressing it asymptotically closer to a representation of instantaneous thought, of an inner flicker of consciousness rather than a practiced performance of speech. That slippage out of social time and into interior time comes to dramatically represent the particular dimension of thought in a way that competes with the more expository (because generally longer) but less formally disturbing dramatic convention of the soliloquy. Hamlet's first utterance thus commences a particular connection between psychological interiority, the maintenance of potentially dangerous secrets, and the possibility of a discourse that both transmits and protects a truth that cannot be shared with the world, and it does so by cultivating a certain kind of attentiveness on the part of the listener and spectator.

Having examined the theoretical implications of the aside, we are now in a better position to return to this first utterance and consider it again:

> *KING:* But now, my cousin Hamlet, and my son–
> *HAMLET:* A little more than kin, and less than kind.
> *KING:* How is it that the clouds still hang on you?[27]

In citing Stanley Wells and Gary Taylor's edition, I have myself selected the most textually conservative option; if the agnosticism of this edition with respect to stage directions that are absent from the original texts does not cancel out the possibility of reading this line as a private aside, it certainly restricts its surface availability, particularly in an edition which is designed to serve as an undergraduate teaching text. Yet an explicit [Aside] notation arguably enforces a nonauthorial construction directly onto the page, instead of in the footnotes, where such encrustations typically adhere. Neither choice feels quite adequate to the incommensurable phenomena of an utterance that can function so successfully as support for two divergent readings. If we smuggle into our encounter with this line the fullness of our prior experience of the play and the character it commences, we can construct equally plausible readings that would shore up either editorial decision.

There are elements of his character (his dispatch of Rosencrantz and Guildenstern, his cold-blooded reaction to the killing of Polonius, his face-to-face confrontations with Claudius under a thin disguise of "mad" speech) which support the reading of this line as a public insult or at the very least as a "stage whispered" pseudo-aside designed to make its point while being done lightly enough to ensure plausible deniability. But there is also a sense in which a reading of this line as a successfully private aside would fit Hamlet's character. The play depicts him from the beginning as prone to displays of linguistic mastery that are somehow also compromised in their successful performance; it shows him already nursing private consoling dreams of public cruelty, however minor. Hamlet plays at participation in the court but also (only) plays at subverting its codes. The breaking of the single letter "d" that converts "kin" to "kind" signals his hybrid status, his "neither/nor"; it is formally indecisive but also aggressive in its display of the refusal to be either term. The pun stages a fantasy of nonrelation to Gertrude and Claudius but aborts it as well; it is precisely in the middle of articulating a relationship to the family bond that language seizes up, and formally ruptures. Hamlet's momentary refusal of kinship here prefigures his later remark to Gertrude: "You are the Queen, your husband's brother's wife, / And would it were not so, you are my mother" (3.4.15–16). It is "unkind" to deny that one is kin, both in its sadistic aggression and in its rhetorical posture as an unnaturally self-sufficient individual. Hamlet longs to de-create himself in order to sever any connection to the mother who bore him, but this desire is obliquely wished for, in a muttered pun. In order to say the word that would be more than kin but not quite successfully reach the end of kind, one would have to catch in the throat, cut off one's speech, stop short. As this reading constructs him, Hamlet's use of a fully private aside would align him with a servant's role, and thus register his own self-understanding as dispossessed of the crown, already playing the part of a disenfranchised and scheming malcontent. According to such a reading, for Hamlet to be openly rude to the King would have political consequences, and so perhaps he cannot afford to utter his line directly to the King; ironically, Hamlet creates or furthers Claudius' power through his own timorous resort to the aside in which he kicks against it.

Hamlet's first line commences a particular dynamic between audience and character. If we understand "melancholy" not as a private trait but as a kind of dynamic relationship of assemblage which solicits interpretation, ascription, and diagnosis in exchange for the teasing revelation of a rhetorically staged interior, then we can trace our relationship to Hamlet as diagnostically activated by our privileged point of access to him. Our capacity as audience members to hear the melancholic aside which others may or may not be able to hear is the first instance of this dynamic entering our experience of the play. That is, the ability to hear the melancholic "aside" which is neither fully public nor fully private constitutes the infra-thin membrane that separates us from the "inside," the "that within which passes show"; it founds a certain interpretive charity toward these potentially empty references by supplying a minimal content for them. It is this experience that supplements the gesture toward the "that within" which allows melancholy to operate in and through us as we complete and ratify its gestures and announcements. If, in absolute epistemological terms, we can never, by definition, have access to the precise contents of "that within," our ability to hear Hamlet's own thought at work *as it reaches out to us* constitutes the high-water mark of our access to him. This returns us to the bare definition of what an "audience" is: "an assembly of listeners," an assemblage whose communal listenership sutures together the bare perceptual activity of listening with judicial and state contexts of trials and hearings and the evaluative and diagnostic dimensions of "audits" and "auditions."[28] Of all the "technical devices" deployed in *Hamlet* (to pursue the Wilsonian reading of what is at work in "the Hamlet mystery"), it is this curious phenomena of overhearing a remark made in public but dramatically marked as private which commences a particular mode of spectatorial experience. Pitting one sense against another in the interests of pitting one individual against a group, the aside formalizes an intimacy between Hamlet and the members of the audience. The melancholic aside constructs an ad hoc perceptual community that dynamically connects Hamlet to his audience while excluding all others, a bond that exemplifies what I have termed Shakespeare's poetics of generative indeterminacy: What remains "indeterminate" is both the nature of the melancholic interiority being kept from view and the precise valence or address of the asides which issue forth from that interiority. In its capacity to interpellate us into service as components of an assemblage and members of an auditory group, the aside mobilizes a potentially unlimited range of meanings and motive-based explanations whose openness remains "generative" (i.e., discursively productive, in circulation, promiscuous).

One might be tempted at this point to object that I am hanging a single quibble about stage directions in rather heavy chains. Does the experience of the melancholic aside, and the collaborative relationship between dramatic character and audience that it establishes, obtain outside of this one line? If the strategic vagueness of the melancholic is productive, what can it be said to produce? What it generates within the world of the play is the interpretive energy of a community organized around Hamlet, an assembled community whose members try to resolve the meaning of his behavior. As Gertrude and Claudius and Polonius soon discover in their attempts to solve the Hamlet mystery in act two, scene two, the problem with trying to interpret melancholy is that the activity works all too well: One can always find some behavior which suggests melancholy, and one can always supply one more speculative causal link which will "show" that it derives from frustrated ambition, excessive grief, spurned love, excessive solitude, excessive religious faith, and so on. There is both an overdetermination of plausible causes for melancholy within the play (Hamlet is in love, Hamlet is in mourning, Hamlet is politically ambitious) and an overdetermination of potentially available definitions of melancholy itself (given the contemporary circulation of conflicting theories about what constituted and caused it).

The interminable analysis of Hamlet's melancholy symptom intersects with the cumulative cycle of misinterpretations that drives the play's broader concern with the problem of intersubjective knowledge. The play depicts a cruel, mutual opacity between Elsinore's vexed, frustrated interpreters: Polonius misinterprets Hamlet, Rosencrantz and Guildenstern misinterpret Hamlet, Hamlet misinterprets Claudius at prayer, and then misinterprets the noise behind the arras, and "thus runs the world away." One might diagnose this steadily increasing catalogue of exegetical error as the play's expression of a strongly skeptical Shakespearean vision of the limited extent of human reasoning, or the absolute inaccessibility of the private self. But in order for such errors in interpretation or rash leaps to judgment to show up as faulty, we as spectators must be in secure possession of a standard against which to measure their failure. In the chapel scene, Claudius confesses to us as spectators that "My words fly up, my thoughts remain below: / Words without thoughts never to heaven go" (3.3.97–98); even as the opacity of Claudius to Hamlet is confirmed, the openness of Claudius' interiority to our dramatic access is established. What remains scandalously unthinkable within this garden variety example of dramatic irony, an *ultima thule* of perversion, is the technically open possibility that Claudius lies to us in his aside, and that he has, in fact, repented; the "impossibility" of such a reading is a sign of the tacit alignment of the aside with truth. The dramatic convention of the aside powerfully connects the hidden with the truthful, and it may be this proto-psychological intuition that has made the play's markedly "interior" moments, such as asides and soliloquies, so irresistible to psychoanalytic critics primed to hunt for the return of the repressed. Given that the dramatic experience of witnessing a misinterpretation requires truth in order to function, the progressive sequence of layered misinterpretations that cumulatively drive the action of acts three and four are themselves demonstrations of the possibility of adequate knowledge, albeit knowledge projected outside the world of the play and into the secure spectatorship of the audience.

The play has proven startlingly elastic in its capacity to generate conflicting accounts of who within the play has seen, and heard, and overheard, what. From John Dover Wilson's suggestion that Hamlet overhears Polonius promising to "loose" his daughter to him (which now has wide critical currency and frequently appears in stage and screen adaptations of the play, as in Franco Zeffirelli's 1990 film) to J. E. Hirsh's suggestion that Hamlet offers his "to be or not to be" soliloquy to the eavesdropping ears of Polonius and Claudius, the possibilities for overhearing and eavesdropping, in short, for promiscuous audition and overlapping but encrypted acts of perception between dramatic characters across all of the play's entrances and exits, has generated a wild profusion of possible readings.[29] Confessed in the midst of its successful suppression, the accessibility of Claudius' guilt to our spectatorship and audition does provide an anchor point of epistemological security, but the security it provides (and, one might add, it is a security afforded by being able to hear Claudius' words rather than see him in an attitude of prayer) is no sooner granted then Shakespeare immediately stages the most disorienting example yet of the untrustworthiness of visual and auditory perception in the uncanny return of the Ghost in the bedroom scene. I turn now to the Ghost's manifestation in the bedroom scene in order to test the persistence of this dramatic dynamic of seeing and hearing "aside" with Hamlet; for this scene allows Shakespeare to further implicate the audience in a direct perceptual experience of Hamlet's melancholy, fueling a kind of preemptive identification with him which will be cumulatively reinforced and extended by the play's conclusion.

Seeing Things and Hearing Voices: Epistemology, Spectatorship, Audition

As audience members, we are presented in the bedroom scene with two fundamentally incompatible yet simultaneous perceptual experiences; we see and hear the Ghost that Hamlet sees and hears, but we also see and hear Gertrude failing to perceive the spectral apparition.[30] If our own access to the Ghost imposes a certain fidelity to Hamlet's perspective, Gertrude's resolute rejection forces us to reconsider, and figures our ability to perceive the ghost as an index of our participation in Hamlet's melancholic madness, thus opening up for us the possibility of a critical disavowal of him: "This is the very coinage of your brain. / This bodiless creation ecstasy / Is very cunning in" (3.4.128–130). Like the experience of the aside, the experience of the ghost vision occupies an interstitial discursive space which is both unverifiable (as it remains unrecognized by the Big Other of the generalized *socius*) and inescapable (we as audience members have only to look at the figure of the ghost upon the stage and we are already brought into the constitutive intimacy of a shared access to an implicating secret).

What are we seeing and hearing when we see and hear the Ghost? The quandary that this moment in the play creates is exemplary of the epistemological problems engendered by the vagueness of the melancholic symptom along a spectrum ranging from moods of depression and malaise to manic displays of aggression to psychotic hallucinations, delusions, and phantasms. In his medical treatise *Touchstone of Complexions* (1581), Levinus Lemnius imagines all forms of "concoction"-based mental illness in terms of a metaphor of an obscure cloud of smoke that compromises clear vision and, by extension, clear thinking:

> If concoction be troubled, there do stryke by into the head, grosse and fumy vapours, such as by example we see greene wood to make, that is smered and covered over with pytche and tallowe. And hereupon it hapeneth that the mynde sometymes conceyveth straunge and absurde imaginations, yea sometimes falleth into dotage, raving, madnesse, phrenzy, melancholy, fure or some other distemperaunce.[31]

The melancholic individual's trustworthiness as a material witness was necessarily undermined by such understandings of their condition as a pathological form of mental and/or physical illness. Read thus, "melancholy epistemology" might signify a dubious state of hallucinatory dotage.

Hamlet's desperation when he overhears Gertrude's "Alas, he's mad" signals the rhetorical loss of persuasive momentum which his access to the Ghost produces. That he feels such desperation on behalf of a truth of which we too are witnesses, also functions as a way of leveraging the audience's emotional investment, of clouding our eyes with the smoke of his pitch and tallow. Seeing and hearing with Hamlet, we are pushed toward an all-too-literal "sympathy" with him (not in the sense of approval but in the original Greek literal meaning of sympathy as "fellow feeling"): Abiding in a state of fellow feeling, we are hardly in a position to decide for or against his own claim to see his father's spirit when we see that spirit too. The act of witnessing and bearing testimony on behalf of "Hamlet's version of events," already begun in the swearing scene, is reinforced here, and such acts of witness will become elevated to the status of a shared ethical task in Hamlet's dying injunction to Horatio to protect Hamlet's "wounded name" by surviving beyond the boundaries of the play and telling his story. In securing a kind of perceptual allegiance with Hamlet through the shared experience of the ghost vision, the play prepares us to share in the melancholy legacy of Horatio's mournful duty to "report me and my cause aright / to the unsatisfied" (5.2.321). In the bedroom scene, it is Gertrude who is unsatisfied, and we as audience members who are grouped into an assemblage as counterwitnesses, poised to verify Hamlet's vision against her denial.

Unsurprisingly, this scene has generated a great deal more exegetical controversy than can possibly be summarized here. It offers an irresistible critical opportunity to shake the possible sources for the play in search of chestnuts of Elizabethan ghostlore, which will reduce the problem to a manageable matter of inheritance. Critical interest in the German play *Der Bestrafte Brudermord* offers a case in point: In that text, the Queen's inability to see the ghost is announced by Hamlet as proof of her murderous complicity with Claudius. Given the divergences across the spiritual literature, folk traditions, and doctrinal disputes about the status of apparitions in the period, recourse to any particular exemplary matter must rest with the provision of probable, but always optional, contexts. What we do know is that in adapting his source, Shakespeare chose not to articulate such a conclusion directly, and instead of folkloric resolution foregrounded the dramatic experience of a corrosive incommensurability instead.

The destabilizing capacity of such perceptions upon those they exclude becomes a portable tactic for Hamlet. Energized by his at least partial "victory" over his mother, who, though she has not seen the Ghost, has at least seen "such black and grained spots / As will not leave their tinct" (3.4.90–91). Hamlet seems to play with, and upon, the power which access to a spirit world unseen by others grants when, in the following scene, he taunts Claudius in a manner which burlesques the bedroom scene's uncannily split perceptions:

> *HAMLET:* For England.
> *KING:* Ay, Hamlet.
> *HAMLET:* Good.
> *KING:* So is it, if thou knew'st our purposes.
> *HAMLET:* I see a cherub that sees them. (4.3.42–46)

Hamlet does not know the precise manner in which a trip to England will further Claudius' ends, but he knows enough to suspect this plan, and to sense the menace in this teasing assertion of his ignorance. Seizing the opportunity to mock and outfence his rival, Hamlet rhetorically marshals a kind of spectral, quasi-participation in the surveillance capabilities available to the heavenly host. Hamlet does not see Claudius' plot, but he reminds his guilty uncle of the existence of divine judgment, even as he seems to mock at its efficacy. The patent ridiculousness of this response, which in its irreligious mockery of divinity reminds us of the aureate, deliberate stiffness of his "Nymph, in thy orisons be all my sins remembered" line to Ophelia, indexes the glib cruelty of Hamlet's melancholy when it is in its manic upswing. Hamlet's vision of the cherub marks the rhetorical use of religious ideas as another mode through which to combat his hated rival by projecting a paternal authority more ultimately powerful than the King of Denmark. But Hamlet's redoubled act of seeing goes beyond this Oedipal skirmish and here again reinscribes the curious dramatic phenomenology of melancholy perception. Hamlet sees an external authority which sits in absolute judgment over the totality of what can be known within the dramatic world of the play; this meta-dramatic intrusion of the melancholic "aside" onto the scene rebounds upon us within the audience, for regardless of what a cherub may or may not know, we in the audience are also, of course, in possession of precisely the knowledge of Claudius' intentions which Hamlet reaches toward here.

"Audience to this Act": Assembling the Witnesses

The importance of securing a faith in the public knowledge of those who spectate upon the melancholic figure is most acute when, after the betrayals and fatal exchanges of the duel scene, Hamlet's death finally approaches. Here spectatorship's constitutive power over the encrypted "truth" of the melancholic figure is transformed from an actively negotiated partnership in the co-creation of "melancholy" to a tragically unilateral relationship of mourning. Fittingly, it is here at the edge of his extermination that Hamlet's mysterious capacity to speak "aside" and solicit not only interpretation but a kind of collaborative fidelity on the part of those who witness him is articulated most explicitly:

> HAMLET: You that look pale, and tremble at this chance,
> That are but mutes or audience to this act,
> Had I but time-as this fell sergeant, Death,
> Is strict in his arrest-O, I could tell you-
> But let it be. (5.2.316–320)

To whom are these lines addressed? Following his ambivalent farewell to Gertrude ("Wretched queen, adieu!," an apostrophe whose courteous French formality jangles with its sympathetic acknowledgment of her suffering but also recalls the Ghost's own warning "adieu" to Hamlet) and preceding his direct address to Horatio, this speech calls forth a set of listeners defined as neither mother nor friend, and constituted by their ability to spectate, but handicapped by their ignorance as to the meaning of what they can see and hear. This founding ignorance requires, but may not receive in time, redress. These lines reach outward, beyond the terminal limits of the court's killing floor, to the spectators before him. Both a flirtation and a foreclosure, Hamlet counterfactually asserts that, were circumstances other than they are, he could tell them—something. Hamlet's tic of self-interruption, established early in the play in sudden swerves such as "There are more things in heaven and earth, Horatio, / than are dreamt of in your philosophy. / But come-" (1.5.166–168), here takes on a new pathos even as it rearticulates the same melancholic strategy. Logically cancelled out even as his modal verb suggests its possibility, this potential speech would not require a reaction, nor would a response in speech even be possible, for we are "but mutes" awaiting a telling that can never quite commence. He has cast his onlookers in a role, and scripts their reactions (pallor, trembling) in a manner that explicitly repeats Ophelia's earlier testimony of his own conformity to melancholic iconography. This speech produces a kind of discursive transmission of melancholy, in effect releasing Hamlet from the burden of his suffering by conveying that suffering to them, and reproducing its symptoms in the therapeutic host-community of onlookers and auditors within the assemblage that this speech names into existence. As such, it is also a pronounced revision of the dynamic commenced by Hamlet earlier in the swearing scene. There he forbids those around him to take up the melancholy repertoire of symptoms; here he actively projects his own condition outward.

There is, of course, a simple sense to the line. The mute audience to which Hamlet refers is the loose set of courtiers, officers, and nobility assembled to watch his fencing match with Laertes. The scene is populated with a nonspecific collection of the Danish nobility and the officers and servants of the King, and this same loose aggregation fills out the "All" who cry "Treason! Treason!" (5.2.305) when Hamlet strikes Claudius.[32] According to this reading, it is presumably to this collection that Hamlet is referring when he identifies a set of onlookers, pale and trembling, who spectate upon the slaughter before them, and it is to this group that he offers the empty promise of a full explanation. But the very fact that Hamlet identifies them as "mute" troubles any easy correspondence between this grouping and the "All" we see upon the stage, for this "All" have been consolidated not by their silence but by their very capacity to unite in vocal denunciation of Hamlet's treason. This act of social cohesion in the passing of judgment for a capital crime (regicide was a crime against both God and the state) unites them in terms of their ability to speak with a single voice, but it would seem to disqualify them as mutes. To be sure, one may stage the scene so that, having loudly denounced the murder of Claudius, they are silenced into respectful sympathy by the spectacle of Hamlet's imminent death, and are only mute within the moment of his dying words. But Hamlet's reference to a "mute audience" here also seems to activate a particular relationship to the dramatic audience outside the play.[33]

The play begins with a portrait of the effects of "maimed rites" upon those who have not been allowed to properly mourn for a death, and in the ghostly emergence of Old Hamlet into the court of Elsinore models the deadly consequences of such a disavowal. Fittingly, it ends with an elaborately manipulative scripting of how we are to respond to Hamlet's own death as it approaches and then overtakes him. Upon closer examination, the response proposed by the play seems delicately poised between the progressive decathexis of normative mourning and the stubborn refusals and introjections of melancholia as outlined by Freud in "Mourning and Melancholia." The royal pomp of a somber militarized funeral promised by Fortinbras provides a posthumous acknowledgment of Hamlet's rightful claim to the throne and the purgation of Claudius' crime, and thus asserts the cyclic rhythm of normative mourning implicit in Claudius' own flat admission "your father lost a father, / That father lost, lost his" and that "this must be so" (1.2.9–90, 106) and so cleanses the "something rotten in the state of Denmark" (1.4.90). Against this current, the final conversation between Hamlet and Horatio works to instill in the audience a kind of secondary melancholia by imposing a duty to preserve an image of Hamlet's just cause against Claudius from the abrasive erasures of a successful, mature, and complete decathexis. In *After Oedipus: Shakespeare in Psychoanalysis*, Julia Reinhard Lupton and Kenneth Reinhard have characterized the link between the play and the reader as part of a

> [. . .] [r]omantic tendency to read Hamlet as doubly autobiographical: Hamlet is Shakespeare, and Shakespeare's modern reader (Goethe, Coleridge, Freud . . .) is Hamlet. This reduplication of autobiography leads not to the transparent, self-confirming identity of reader, text, and author, but to an allegorical splitting that infects interpreting subject, literary object, and authorial ground. Hamlet's legacy is a layered identification with the grieving son who has himself infolded the murdered father.[34]

This expanded sense of the identifications made available by the play both between its characters and between the play and reader is crucial because it links the dynamic of transmission, preservation, and fidelity that obtains between Hamlet and Old Hamlet with the final "legacy" that the dying Hamlet imposes upon his "mute and trembling" audience—both within and without the play. It proposes a flexible series of relationships of identification (always partial, and always achieved against some resistance from within the play's capaciously suggestive poetics) which can, in the transmission of Hamlet's "cause" from Hamlet and Horatio to the audience, also model the linkages between the Freudian understanding of Hamlet's melancholia and the wider intersubjective assemblage of dramatic melancholy that *Hamlet* allows us to experience and complete.

Hamlet's own death and the carefully secured conditions under which he will be mourned and remembered completes the process of "layered identification" and blurring between himself and the Ghost of his dead father (a process which seems to have begun immediately upon his father's death, if we are to believe Claudius and Gertrude's references to his obstinate mourning which extend proleptically before the action of the play). Hamlet's extended death speeches conclude with a telling image of the transformations taking place inside his body: "O, I die, Horatio / The potent poison quite o'ercrows my spirit" (5.2.334–335). In saying this, Hamlet places himself in the precise position of his own undead father who speaks and describes his own poisoning with a speech that comes from the other side of death. But in doing so, Hamlet comes to resolve the tension between his melancholy and the commandment to act; he is discharged of his father's imperative and his life simultaneously by occupying the place prepared for him by each of his two fathers, Old Hamlet and Claudius.[35]

Hamlet's commanding injunction to Horatio against suicide would seem to suggest that he himself has gone beyond the suicidal urges and aggressions which held sway over him in the "to be or not to be" speech, when agnostic ignorance of the afterlife seemed to be the only factor which held his self-destructive energies in check. But this capacity to issue such commands must also be read as part of the play's ongoing broader consolidation of Hamlet as a figure who becomes increasingly regal as he approaches death, a dramatic curve which reaches its apogee in Fortinbras' posthumous acknowledgment that he was likely to "have prov'd most royal" (5.2.380). Indeed, the stakes of establishing and verifying masculinity, and of locating Hamlet in relation to available modes of masculinity, become acutely present at the moment of his death. It is as if the asymptotic approach to the absolute passivity of a fallen corpse triggers a vigorously active production of regal imperatives and war-like commands as he negotiates Horatio's mourning duties and hierarchizes the homosocial bond of intimate, loving friendship between them by forbidding Horatio's self-destructive impulse and conscripting him into the duties of a posthumous herald:

HAMLET: Horatio, I am dead
Thou livest. Report me and my cause aright
To the unsatisfied.
HORATIO: Never believe it;
I am more an antique Roman than a Dane.
Here's some liquor left. (5.2.320–325)

This reference to antique Roman culture resonates with the allegation that this play mobilizes a set of references designed "to please the wiser sort," insofar as it obliquely references Senecan stoicism.[36] Horatio, as a scholar, can make reference to Seneca's heroic acceptance of honorable suicide over survival in disgrace by simply calling himself a "Roman," with this designation metonymically standing in for both a specific philosophical tradition within that culture and, more specifically still, with its most celebrated act of practical application. The easy referential familiarity with which Horatio can telegraph his intentions to Hamlet through this hallucinatory claim of national allegiance to ancient Rome over contemporary Denmark models the kind of scholarly masculine sodality of urbane intellectual humanist elites within certain court circles in London. But the repudiation of a Senecan death in favor of survival marks a tense limit to the availability of such models of behavior.

Faced with Horatio's pledge of death, Hamlet rebukes him, and his rebuke makes clear the ambitious stakes of this refusal of solidarity to both his friend and to the "mute and trembling" assembly who stand and spectate upon his final breaths:

HAMLET: As th' art a man,
Give me the cup. Let go! By heaven, I'll ha't!
O God, Horatio, what a wounded name,
Things standing thus unknown, shall I leave behind me!
If thou didst ever hold me in thy heart,
Absent thee from felicity awhile,
And in this harsh world draw thy breath in pain
To tell my story. (5.2.326–331)

Suicide was a mortal sin. But Hamlet's repudiation of Roman suicide does not advocate for a Christian heroic of long-suffering passivity so much as it relocates stoic values in an ongoing duty of painful survival over a temporary and dramatically overdetermined act of momentary violence. Note Hamlet's reinscription of gendered identification as an alternative to national identification. Horatio has asserted his Roman-ness over his Danish-ness; Hamlet responds by asserting first and foremost his masculinity as a primary quality which precedes, and trumps, national identification as such.[37] The legacy of Hamlet is the legacy of discourse, the stable maintenance of an account of what has been seen which will abolish the skeptical threat of "things standing thus unknown." This is a heavy burden indeed, for Hamlet has himself abided in precisely the unknown fissure opened up between what he can see and hear and what he can know or verify as he struggled with the testimony of the ghost and the troubling reality of his mother's sexuality. For us within the audience, much of what we would need to transmit in order to rescue Hamlet's name from "wounding" itself abides in a corresponding "unknown" opened up by Hamlet's ceaseless, yet empty, gestures toward a space of melancholy subjectivity which he has actively constructed as outside of our grasp. Hamlet seems familiar with the affect that accompanies such a task, but he refers to it with a characteristically oblique negation: It is the absence of felicity. In other words, Hamlet's legacy is melancholy, melancholy understood now in its final form as the condition of an assemblage of witnesses who must mourn, identify with, encrypt, and guarantee the now-vanished traces of their intimacy with an "unknown."

5. *Rhapsodies of Rags*

A Melancholy Stereoscope

Imagine a book. This book fixates upon a particular cultural phenomenon, one not entirely marginal but far from obvious as a suitable subject for its imposing size. The book drifts from its stated occasion, relentlessly, and uses the explanation of its ostensibly modest topic to digress at great length across the entirety of the surrounding world, pursued to its stoppages and endpoints in metaphysical speculations and skeptical aporias. At each local point in the book, the overarching theme is sounded, and yet the very multitude of examples of this motif proliferate to such an extent that, gradually, one begins to suspect that the author may have monomaniacally imposed a connection upon anything and everything he chanced to encounter in his scholarly researches. The book's structure is rigorously foregrounded by an exoskeleton of curiously organized thematic divisions and subdivisions, and yet this structure is promiscuously open enough to interpretation that one might justifiably locate the same exemplary subsection under countless other headings with just as much sureness of fit. The book absorbed its author utterly, occupying him for decades at the expense of any other productive labor, for its peculiar composition demands an oceanic accumulation of significant details, requiring countless hours in libraries amassing evidence, counter-evidence, proof, examples, new citations, and variant translations of old citations, in a quest for the raw material out of which to fashion the work. These quotations constitute the overwhelming majority of the book itself, dissolving the boundary between archival compilation and creation, and forcing its readers to wonder who its actual author is, and what this obsessive compiler intended. At once too much and not enough (for the book could always absorb more matter), the book consumes the author's life; from the moment he begins to write the book, he is occupied with it until his death. The book constitutes a kind of encyclopedic reconstruction of the surrounding world, which fashions out of its magpie hoard of scholarship a critical reaction to the society from which its author has studiously retreated. More referred to than read, the book is often started and rarely finished; it is too shapeless in its form and cloudy in its intentions to hold a casual reader. The book is now the preserve of a small community of bickering experts.

There are two books which fit this narrative: Robert Burton's *The Anatomy of Melancholy* and Walter Benjamin's *The Arcades Project*.[1] I must wager this claim of kinship against the obvious fact of their substantial difference: Burton takes as his subject the dwindlingly fashionable humoral illness of melancholy, while Benjamin tackles the first Parisian glass-ceilinged shopping arcades (commonly, if misleadingly, regarded as the predecessors to today's shopping malls). Furthermore, these two texts remain marked by wildly distinct histories of reception: For a quasi-medical tract of consolation, Burton's text was a runaway popular success, and the lifetime of revision which led Burton to produce six different editions before his death was spurred as much by market demand as personal obsession. By contrast, Benjamin's text took a darker path, existing in manuscript as a perpetually reshuffled collection of notes for a "final" version that remained unwritten at the time of his suicide. One exists as six different books; the other arguably does not exist at all. And yet, as experiences for the reader today, these texts use dramatically similar formal methods and produce an uncannily similar effect. Burton's seventeenth-century *cento* of self-canceling fragments of scholarly lore and Benjamin's twentieth-century ruined anthology of telling fragments of nineteenth-century cultural history both function as creative and personal assemblages expressive of what I wish to call "melancholy structure."

The explicit framework of the *Anatomy*'s exoskeletal "table" foregrounds its ambitions as a medically useful self-help book; the *Anatomy*'s "Partitions, Sections, Members, and Subsections" create a textual body whose divisions arrange an encyclopedic scope of material into localized zones so that sufferers from particular types of melancholy can check their symptoms against Burton's lists, or can proceed swiftly to the available

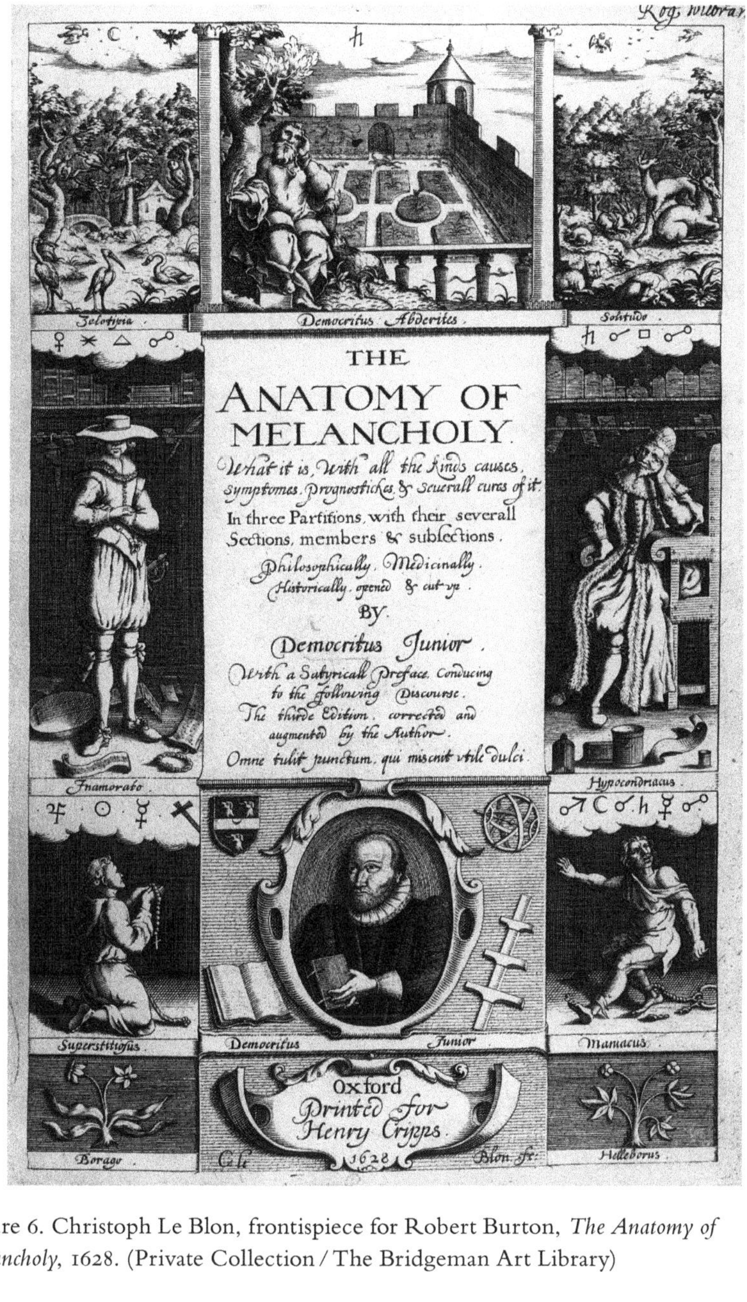

Figure 6. Christoph Le Blon, frontispiece for Robert Burton, *The Anatomy of Melancholy*, 1628. (Private Collection / The Bridgeman Art Library)

cures. Yet the actual text contained within the forms set out in those tables undermines the reader's confidence in the usefulness of his advice, and the relationship between therapeutic plan and skeptical tactics grows increasingly tense as the text progresses into the second and third partitions. That is, the reader proceeds from trust to doubt as the exemplary material increases—what is generated above all is a sense of delight in the onrushing procession of cultural materials, shot through with a nagging undercurrent of unease with the dwindling prospects of there being any activity which is not a possible expression of melancholy, or a remedy which has not been recommended by one authority but repudiated by another. Benjamin's structure is more atomized and peculiar: The *Arcades Project* exists as a set of entries and research notes organized under thematic headings that he termed *Konvoluts*. Some examples of *Konvolut* headings include: Fashion, Boredom, The Collector, Panorama, Mirror, Doll/Automaton, Prostitution, and Iron Construction, among others.[2] Benjamin's text gathers its exemplary material under these headings and oscillates between simple citations of sources of cultural history and a kind of critical caulk or connective tissue which points us toward what in particular he found telling or revelatory about each citation. This enormous database of material on the climate of bourgeois capitalist Paris seems designed to trigger a mingled sensation of dreamy admiration and Marxist disdain; aphoristic pronouncements and occasional critical paragraphs interrupt long stretches of reverie amongst fragmented arcana.

In both texts, the extended temporality of ceaseless collecting and antihierarchical, nonsystematic network-building which their authors engage in induces in the reader a particular affective state: Both texts encourage a passive, browsing flow, a free-floating suspension of judgment, a kind of scholarly languor in the gaze upon a particular unit of "raw matter," such as a passage from Laurentius or a Parisian advertisement. This psychic state of pleasurable reverie, in which the reader is weighted down by an experience of numbing heaviness that could seemingly extend forever, tellingly replicates the symptomatology of genial melancholy. An experience of aesthetic contamination takes place which activates the transmission of knowledge into something closer to direct experience: The text that describes the sluggish yet pleasant tranquility of melancholic contemplation itself requires that the reader enter into that state in order to move across its vast, gnarled surface. Gradually, one seems to get lost in the accumulating flow of particular details. This sensation of "endlessness" obtains at each particular moment in the work, because one could always supply more citations to the end of the "&cs" that stud Burton's prose, and because one could always supplement the *Arcades Project* with one more anecdote, or one more *Konvolut.* Yet it also obtains globally, across the reader's experience of reading these works, due to their massive bulk: The most current scholarly translation of the *Arcades Project* runs to 1072 pages, while the Oxford edition of the *Anatomy,* excluding indexes and commentary, runs to 1148 pages.

Both Benjamin's and Burton's texts ceaselessly shuttle between the representation of an oceanic "outside" (the fragmented archive of exotic found materials) and the production of a corresponding psychological/somatic "inside" (a virtual interior constituted by the scholar's direct address to the reader, or the scholar's address to himself, which both transmits and models melancholy as a state of being that is both embodied and psychological). Burton's introduction to the subsection on religious melancholy models this exemplary shift from the "oceanic" to the "interior":

> Give me but a little leave, and I will set before your eyes in briefe, a stupend, vast, infinite Ocean of incredible madnesse and folly: a Sea full of shelves and rockes, sands, gulfes, Euripes and contrary tides, full of fearefull monsters, uncouth shapes, roring waves, tempests, and Siren calmes, Halcyonian seas; unspeakable misery, such Comedies and Tragoedies, such absurd and ridiculous, feral and lamentable fitts, that I knowe not whether they are more to be pitied or derided, or may be believed, but that we daily see the same still practiced in our dayes, fresh examples, *nova novitia,* fresh objects, of misery and madnesse in this kinde that are still represented unto us, abroad, at home, in the midst of us, in our bosomes.[3]

Conjuring up a beguiling seascape of madness for the reader, Burton here affects the tone of an omnipotent narrator, only to stop short with a telltale self-sabotaging admission: His "I know not" suspends the reader uncomfortably between therapeutic seriousness and satirical mockery, preparing us for the final twist in which what seemed to be an exterior landscape of other people's madness becomes a mirror held up to the folly lodged "in our bosomes." By the end of the sentence, not only do we not trust Burton but we do not trust ourselves. Such a text is and is not in the business of imparting knowledge; while the vastness of its textual materials and explanatory scope promises to make monstrous erudition available to us in a portable, encyclopedic form, its skeptical resistances and persistent unreliability call the epistemological foundation of a guidebook (medical, religious, philosophical) into question. From start to finish, Burton's and Benjamin's texts are both dauntingly massive in scale yet curiously flippant and "bemused" in tone, and this mingled heaviness-in-lightness, or glancing, capricious shifts between tonal registers, defines the particular spell that both works cast upon their readers.

In *Konvolut* N, titled "On the Theory of Knowledge, Theory of Progress," Benjamin describes his technique: "Method of this project: literary montage. I needn't *say* anything. Merely show. I shall purloin no valuables, appropriate no ingenious formulations. But the rags, the refuse—these I will not inventory but allow, in the only way possible, to come into their own: by making use of them."[4] Benjamin's phrase strikingly repeats in a somewhat softer key an image which appears in Burton's apology for his crude structure and style in the prefatory "Democritus to the Reader":

> And for those other faults of Barbarisme, Doricke dialect, Extemporanean stile, Tautologies, Apish imitation, *a Rapsody of Rags gathered together from several dunghills*, excrements of Authors, toies and fopperies, confusedly tumbled out, without Art, Invention, Judgement, Wit, Learning, raw, rude, phantasticall, absurd, insolent, indiscreet, ill-composed, ill-digested, vaine, scurrile, idle, dull, and dry; I confesse all ('tis partly affected) thou canst not thinke worse of me than I doe of my selfe.[5]

Both texts engage in a relentless appropriation and framing of found materials, and both refrain from a directly polemical argument about how we should respond to those materials—defiantly flinging them at the reader with a deliberately foregrounded refusal to explicitly comment upon them and resolve or explain away the tension generated by their presence upon the page. Yet, paradoxically, despite this stated desire to simply pile rags and excrement before the reader without explanation, each text pulses with authorial energy beneath the surface, constantly soliciting a kind of oblique judgment on our part about each particular item being cited. This half-submerged, liminal authorial presence in which the scholarly author seems to hint at a stance toward the object of his attention while remaining veiled and opaque, constitutes a new mode of melancholy performativity.

If the dramatic spectacle of melancholy involved an endlessly productive yet aporetic dynamic between the stage melancholic and their puzzled interpreters (both within and without the play), the scholarly equivalent within these texts involves an endlessly additive yet inconclusive accumulation of "raw material" occasionally interrupted by the entrance of a scholar figure whose own manner participates in both poles of the dramatic split: Both Burton and Benjamin seem equally at home playing puzzled interpreters and melancholic wanderers. In part, their aggressive vomiting forth of pointedly chosen snippets and fragments of popular culture recalls the canny thrust of dramatic "mad speech" (think of Ophelia's catches of popular songs or Hamlet's "pregnant" excerpts from the "words words words" he reads). Burton's Democritus persona and, to a lesser extent, Benjamin's own stance as Virgilian guide through the inferno of the nineteenth century's remains, are marked by a curious oscillation between occasional outbursts of critical venom (messianic Marxist forecasts of annihilation from Benjamin, satirical jeremiads from Burton) and long stretches of muffled reticence about their attitudes toward the materials they assemble. If we are occasionally buttonholed, we are more often abandoned to swim through their archives for ourselves, but this deliberate retreat sauces the inevitable return of authorial intrusion. Triggering the reader's longing and building up an intellectual form of suspense, this strategic inconsistency of tone transposes the interpretive dynamic of dramatic representations of melancholy into the generic confines of academic prose. The laborious effort and obsessive absorption in the gathering of quotations which the author undergoes in order to create his work is reexperienced by their readers as a pilgrimage across an intellectual landscape of artfully reconfigured scraps and partial objects. Each author intercuts their vast necropolis of "dead" texts with jolts of contemporaneity: the latest contributions in scientific and medical literature and religious controversy in Burton's case, and the latest philosophical and critical developments in Benjamin's case. We are offered the promise of a deeper knowledge of our "subject" (melancholy, arcades) through an immersion in its fragments and traces; yet the syntax of these fragments seems to inadvertently call the stable identity of that subject—its capacity to generate knowledge at all—into question even as a secondary, occluded "subject" (the author himself) alternately glances into view and resists us. Both texts produce a kind of delight at the procession of charmed arcana put on display, and yet this unpredictable *dérive* is shot through with an undertow of mournful, saturnine heaviness.

Both Burton and Benjamin famously linked themselves to Saturn, the planetary lord of melancholy. Early in his preface Burton, a keen astrologer, announces grandly that "Saturn was lord of my geniture,"[6] and Benjamin, mingling pride with dramatic foreboding, echoes him with an added gloss for the modern reader: "I came into the world under the sign of Saturn—the star of the slowest revolution, the planet of detours and delays."[7] A few pages after declaring the saturnine provenance of his birth, Burton directly identifies himself as a sufferer of the very disease he is anatomizing:

> I was not a little offended with this maladie, shall I say my Mistris Melancholy, my Aegeria, or my malus Genius, & for that cause as he that is stung with a Scorpion, I would expell *clavum clavo*, comfort one sorrow with another, idlenes with idlenes, *ut ex vipera Theriacum*, make an Antidote out of that which was the prime cause of my disease.[8]

Through such melancholy self-portraiture, Burton bequeathed to posterity a suitably stylized image of himself: "the fantastic great old man," as Charles Lamb described him, a high priest of melancholy consumed by his own scholarship, lost in the libraries of Oxford.[9] Benjamin shares equally in this project of self-melancholizing. At the beginning of his *Konvolut* on the theory of knowledge, Benjamin pauses to describe the environment in which he writes, and in the process depicts himself as the melancholic scholar of modernity subtly taking on the frozen stillness of the *trompe l'oeil* that decorates his surroundings:

> These notes devoted to the Paris arcades were begun under an open sky of cloudless blue that arched above the foliage; and yet—owing to the millions of leaves that were visited by the fresh breeze of diligence, the stertorous breath of the researcher, the storm of youthful zeal, and the idle wind of curiosity—they've been covered by the dust of centuries. For the painted sky of summer that looks down from the arcades in the reading room of the Bibliotheque Nationale in Paris has spread out over them its dreamy, unlit ceiling.[10]

As with Burton, "idleness" is revealed to be the melancholic lining of the vibrant display of research and archival activity that produces the scholar's text, the uncanny twin of its busyness. The eternal, and eternally dead, summer sky of the library's painted ceiling has here lent Benjamin's own historical excavations, carried out beneath them the paradoxical "timeless mortality" of a *nature morte*. Describing the heraldic power of the image of the melancholic as an augur of a veiled, mysterious knowledge, Benjamin wrote that early modern individuals "were able to recognize in the features of the sorrowful Contemplator the reflection of a distant light, shining back from the depths of self-absorption."[11] Viewed stereoscopically, these paired portraits of Burton and Benjamin unite to form a single image of melancholy knowledge, of melancholy as both the subject and the result of scholarship, with the pursuit of knowledge displayed for the reader as one which "increaseth sorrow" while also generating an intoxicating pleasure, allowing these scholars to luxuriate in a particularly aestheticized, and remarkably productive, mode of cultivated melancholic suffering. But these images come to us only as themselves a reflection cast by the flickering starlight of their endless, and endlessly generative, melancholy texts. We cannot know who lies inside them.

If the alternately elegiac and savage authorial voices of Burton and Benjamin as scholars present us with a new form of melancholic self fashioning, the structural arrangement of their texts out of partially preserved fragments also models a new melancholic relationship to the object of knowledge. At the level of tone, the shift from the fatal machinations of Shakespeare's *Hamlet* to the satirical accumulations of Burton's *Anatomy* bears out Marx's dictum that tragedy returns as farce. But there is more than a return to comic modality in play, for in this process of transposition from drama to prose melancholy's meaning undergoes a kind of universalizing apotheosis; no longer the symptom of a particular imbalance or individual temperament, Burton's ability to discover melancholy in everything and to present everything as a potential cause or cure for melancholy both expands and nullifies his subject. In this sense, Burton's *Anatomy* marks both the highwater mark and the vanishing point of a certain historical trajectory. That said, the particularly antihistorical structural tactics of Burton's melancholy archive remain an open source for others, notably, I shall argue, Walter Benjamin. Taken together, the kindred assemblages of Burton and Benjamin offer an affectively immanent experience of melancholy as form, modeling what, with apologies to Raymond Williams, we might term the feeling of structure.[12]

Burton Criticism as Melancholy Ruin

Such a celebratory tone might well seem out of place within the bunker/crypt of Burton criticism. Contemporary critical engagement with *The Anatomy of Melancholy* has not entirely recovered from Stanley Fish's assault upon Burton in *Self-Consuming Artifacts*. Frogmarching the reader toward a description of the *Anatomy* as a skeptical abyss, Fish's reading reaches programmatically Fishian conclusions with great vigor: Burton's text is "a series of false promises which alternately discomfort the reader and lead him on."[13] Alternating between exasperation at its frequent switchbacks and admiration for its relentless energy, Fish claims that the *Anatomy* reduces melancholy to a merely occasional opportunity for the staging of the impossibility of knowledge (both clinical and textual). In doing so, Burton allegedly acquits himself of any authorial accountability for his therapeutic task by fashioning only "distracting confusion," a predictable, and predictably "Fishian," result which nonetheless does genuinely capture much of the *Anatomy*'s dislocating effect upon the reader.[14] Though he presents a careful account of his own experience as a reader whose hopes of airtight argumentative structure are dashed on the rocky shores of Burton's prose, Fish shows a surprising lack of interest in considering whether Burton might be working to produce an intentional frustration of his generic and formal expectations about how prose arguments should be structured and how moral questions should be resolved, or of how that frustration might be conceptually related to the specific epistemological distortions of melancholy. When held against the moral standard of "good physicians" such as Milton and Plato, Burton seems, according to Fish, worryingly agnostic about the prospects of improving our minds, an admittedly sour conclusion to reach at the close of an enormous text ostensibly dedicated to precisely that. Ultimately, Fish's argument does not suggest that we ourselves need to read Burton carefully or closely but seems to function rather as a reassuring explanation of why we need not bother moving much past the preface.

If Fish's own interests have since moved on, Burton's critical estimation has not, for Fish's text remains all too typical of how, and in particular where, people tend to read him. Fish wisely sticks to the preface to support his Democritus-oriented thesis, and in large part both those who agree and those who disagree with Fish have followed suit by either endorsing or disputing him through their own responses to the preface, thus allowing the vast, forbidding expanse of all three primary partitions to gather cobwebs when it might more productively be mined for counterexamples. Unfortunately, Burton criticism in the wake of Fish has primarily been concerned with improving his image by stressing Burton's sincerely therapeutic goals, either through a renewed emphasis upon his position in the Church (Michael O' Connell's *Robert Burton* and E. Vicari's *The View From Minerva's Tower*), through a reassertion of the clarity of structure throughout the *Anatomy* (Ruth Fox's *The Tangled Chain*), or through a renewed emphasis upon Burton's participation in a traditional generic form (Mary Schmelzer's *'Tis All One*). These readings have been, in Ruth Fox's telling phrase, "more exhaustive if less compelling" than Fish's essay.[15] More recently, Mary Ann Lund's *Melancholy, Medicine and Religion in Early Modern England: Reading The Anatomy of Melancholy* (2010) has strongly repudiated Fish's reading of Burton, instead reasserting a therapeutically engaged and "serious" Burton through an assessment of his variable rhetorical strategies and a new consideration of the text as a sequence of curative acts.[16] As responses to Fish, all of these texts seem in practice to grant Fish's self-serving distinction between "explanation" and "taxonomy" (the former is his approach, the latter, the merely accurate works of modest intellectual history produced by others). I wish to suggest that the cumulative experience of the entirety of the *Anatomy* generates something richer and stranger than Fish's alternately bored and exasperated skepticism and, in particular, that Burton's complex palimpsest of endless quotation, self-satirizing mockery, and lifelong revision provides an original and forward-thinking model for how to perform scholarly work which is both slyer and more generative than either Fish or his respondents let on.

For better or for worse, Fish's text has so far succeeded in polarizing Burton scholarship into a defensive posture in which the unifying critical question of the past several decades remains one of authorial intention: Is Burton serious or not? Neither Fish's "no" nor his respondent's "yes" permit us to engage the extent to which Burton's text formally enacts something other than intentionality-drenched authorship in the first place. If we transpose the question, we can see the desubjectivizing and estranging force of the assemblage at work: If it makes little sense to ask "is this transpersonal archive of quotations serious or not?," this only flags the need to think otherwise: How is this transpersonal archive of quotations organized? What kinds of connections and experiences does that organization permit? How does melancholy as an object of knowledge emerge or recede from accessibility within this text? What are the cumulative effects of its unself-sameness over time and across the material expanse of the text?

Burton's *Anatomy* arguably occupies the summit of a particularly rich early modern intellectual trajectory: Through its encyclopedic reach, definitive status, and persistent popularity, it sums up and completes the transmission of the melancholy archive from classical texts through its Ficinian continental renovation into the examining rooms of physicians and onto the public stages of English popular culture, at once crowning and closing a tradition. Simply put, it is hard to imagine how it could accomplish all of this if it was merely a frustrating scholarly shaggy dog story or the protracted prank of a Christ Church crank. As it much as it looks backward, Burton's practice of textual assemblage also powerfully anticipates the future, putting in place an original structural relationship between melancholy and knowledge whose dynamic strategies and peculiar effects persist in the untimely cultural histories and unfinished archives of modernity's recording angel, Walter Benjamin.

To risk the claim that Burton originates anything at all, one must first reckon with his debts. The idea of a satirical anatomy is not original to Burton; Philip Stubbe's *Anatomie of Abuses* (1583) is simply the most familiar of a wide range of early modern textual anatomies. Furthermore, the anatomy of the *Anatomy* has a long pedigree, as Burton himself is quick to acknowledge: "For my part I have honorable Presidents for this which I have done: I will cite one for all, *Anthony Zara Pap. Episc.* his *Anatomie of Wit*, in foure Sections, Members, Subsections, &c. to be read in our Libraries."[17] So it is not particularly revealing to point out the overall structure of *The Anatomy of Melancholy* and assert that Burton's innovations are "structural," if by this we simply mean the division and subdivision of a general topic. It must also be said at the outset that the construction of a compendium of thematically organized quotations is, of course, not specific to Burton either; Burton is keen to locate the *Anatomy* within the territory of Italian *centos* and English commonplace books ("I have laboriously collected this *Cento* out of divers Writers").[18] Yet this awareness of his self-conscious participation in the *cento* tradition is balanced by a certain pride in the particular manner in which he has constructed the *Anatomy*, for even though "we can say nothing but what hath beene said, the composition and method is oures only, and shewes a Schollar."[19] Aware that he is making a *cento* as others have, and that he is not saying anything which has not already been said, he nonetheless insists that his "method" in doing so is his resolutely his own: What can this surprising assertion mean?

Generically speaking, *The Anatomy of Melancholy* simply takes up certain ready-to-hand authorial practices: the *cento* or commonplace book as a new textual body fashioned out of borrowed parts, Menippean satire as a genre which licensed social critique of contemporary mores, the utopia as a thought experiment with an estranging force upon the hold of the present, and the anatomy as a punning metaphorical hinge between somatic and discursive registers.[20] Furthermore, the vast cultural archive of Hippocratic medicine, Galenic humoral writings, the pseudo-Aristotelian *Problemata*, the genial skepticism and oblique self-exposition practiced by Montaigne in his recently translated *Essays* (1592), and Ficino's ambitious neo-Platonic reading of melancholy establish an extensive territory of melancholy knowledge and literary models, providing Burton with a richly conflicted textual heritage ripe for scholarly dissection. Of course, the range of Burton's "raw material" extends beyond his titular topic. Widening the explanatory scope of his text further, Burton's all-consuming maw searched with equal hunger for examples of melancholy within all modes of human behavior and forms of knowledge, and raided any available field of inquiry (or at least all that could be found within the texts of the Bodleian library) for diverting tales which might provide distraction from melancholy's effects. Resembling the baggily inclusive catalogue sentences offered up by object oriented ontology, Lawrence Babb's wonderfully Latour-esque list telegraphs the diversity of Burton's moldering hoard and suggests that Burton already anticipated this collage-like turn toward disjunctive inventory:

> One reads of alchemy, Rosicrucianism, comets, the mines of the New World, aphrodisiacs, heraldry, botanical gardens, magnetism, chiromancy, Queen Elizabeth, Socinianism, Helen of Troy, the misanthropy of Timon of Athens, falconry, tobacco, the evanescence of beauty, the venality of lawyers, amulets, simony, the vicissitudes of fortune, the decay of the world, Arminianism, Mark Anthony and Cleopatra, miraculous

> cures, the sin of curiosity, *praestigiae daemonum*, sycophancy in prince's courts, the machinations of the Jesuits, etc. etc.[21]

Sufferers from melancholy were already depicted as prone to both monastic retreat from the world and aggressive critique of those around them, and so this melancholic archive also provided Burton with an enabling satirical mask in the figure of the melancholic, mocking hermit-philosopher Democritus. Burton's own methodological innovation, what remains "his own" in the midst of all of these borrowings and inheritances, was his perception of the synthetic, creative possibilities in the application of the anatomy/*cento* form to the melancholic archive. For the specific intellectual contours of melancholy as a generatively indeterminate node between conflicting discourses and understandings (melancholy as substance, as temperament, as illness, as gift) proved uniquely resonant with, and transformative of, the open-ended, nonhierarchical, antirhetorical, discursively flexible *cento* form. That is, Burton's innovation, if it is appropriate to use a term so invested with modernist aesthetic values, consists not simply in a canny act of noticing ambient modes. Burton did not simply survey the available literature and recognize that melancholy had finally become a capacious enough topic that it would be suitable for an encyclopedically expansive text; *he made it capacious* through his creative acts of expansion and connection.

But Burton's particular achievement goes beyond the quantitative novelty of building something vast out of a slender topic.[22] Rather, it consists in creating a particularly dynamic interplay between these specific structures (satirical anatomy, medical anatomy, *cento*/commonplace book) and a specific subject (melancholy at the crossroads of a conflicted heritage, melancholy as sign and subject of scholarship). Though this particular dynamic is best experienced on the ground in the minutiae of each subsection and sentence, the following broad patterns of thought could be said to characterize the *Anatomy*:

1. Burton treats history as panoramic rather than cumulative; he *surveys* history as a total set of available ideas rather than as a narrative arc of emergences and disappearances; he is in this sense "antihistorical."
2. Burton seeks to distract rather than persuade; treating the digression *as* the rhetorical set piece, *The Anatomy of Melancholy* replaces argument with assemblage as the immanent experience of textuality.

3. Burton divorces citation from endorsement; by keeping the distance between quoting a text and supporting its claims endlessly variable and open, Burton cites others not to reinforce his own authority but to keep authority itself permanently in question.

It is these particular intellectual habits and strategies that I shall argue suggest a kinship, an uncanny correspondence in both method and matter, between Robert Burton and Walter Benjamin. It is a curiously perverse kinship, a shared syntax of citation through which authorship itself is sidestepped in favor of more porous compositional strategies, which one might compare with wiki-esque archive-building, quilt-making, audio sampling, or "patching" on a modular synthesizer. Across religious and political differences and a distance of centuries, the two scholars are united in their methodological practice of responding to culture through the construction of fragmented melancholy assemblages that level down and drift across a historical range of raw material. Each engages in a powerfully seductive mythologizing of their own identity as a scholar who both studies melancholy and simultaneously models it for their reader, generating an experience of the very fragmentation and reverie that their work describes. I intend to argue that Benjamin's writings on melancholy can be brought into a productive relationship with his writings on historical methodology, and that doing so can allow us to reconsider the dynamic interaction between Burton's fragmented praxis and his melancholic subject. But first I wish to clarify Burton's achievements before advancing to his successor and melancholic twin.

Burton does not contribute new information to the archive. Rather, he performs a creative act of scholarship upon it, building new networks, cross-indexing its historical components along new thematic axes. As he does so, he exceeds (and, to give Fish his due, perverts) his explanatory occasion through a pronounced tendency to dissolve differences, to signal and then blur distinctions, to disappear down blind alleys, to lose his way in the labyrinth. These strategies and techniques constitute Burton's own methodologically original response to the archive of melancholy: He rearranges preexisting materials in a new way that both dissects and enacts melancholy for the reader, and encourages an active participation on the part of the reader in the author's (dis)associative logic, a hermeneutic of affective contamination. Pursuing Burton's train of thought requires a kind of collaborative assent in the endless flow of reference and counter-reference, a surrender to the work's dilatory rhythmic pulse. As the reader witnesses the author flattening, disordering, distracting, and hiding within these found materials, these strategies and techniques exhibit melancholy's effects upon a working scholar. That is, Burton is not only describing melancholy through the assembled texts of the *Anatomy*; as Democritus Junior he is also performing melancholy for the reader and inviting the reader to enter into a *folie a deux* of absorption, withdrawal, and obsession with him. It is not simply a work about melancholy but a work made by "melancholizing." As Jonathan Flatley notes in *Affective Mapping*, "Burton seems less concerned to offer a coherent account of melancholy than to find ways to be able to keep writing about it."[23] This foregrounding of writing as process and activity through which to manifest melancholy is confirmed by Adam Kitzes' observation that "the suspicion that Burton's text is about expansion may say more about melancholy than anything within the text itself."[24] His shifts and feints as an author revisit upon the reader his own description of the condition at hand: "as a man that's bitten with fleas, or that cannot sleep, turns to and fro in his bed, their restless minds are tossed and vary."[25] Burton's celebrated defense of his style naturalizes into a landscape this erratic manner of writing:

> 'tis not my study or intent to compose neatly, which an Orator requires, but to expresse my selfe readily and plainely as it happens. So that as a River runnes sometimes precipitate and swift, then dull and slow; now direct, then *per ambages*; now deepe, then shallow; now muddy, then cleare; now broad, then narrow; doth my stile flow: now serious, then light; now Comicall, then Satyricall; now more elaborate, then remisse, as the present subject required, or as at that time I was affected.[26]

The tension between Burton's pledge to express himself "readily and plainely" and the absolute license which his river metaphor provides him glimmers at us beneath the smoothly naturalizing flow of its own description; as the phrases burble and bounce across the page, we are entrained into a kind of synchrony of assent with them as a pleasing pattern even as they announce to us the unpredictability and unreliability of the voice that speaks to us.

If this restlessness expresses melancholy at the level of his tone and syntax, his methodological aversion to chronological order conveys it at the level of ideas. The antihistorical tenor of Burton's engagement with the temporality of his sources makes possible for the reader a kind of subtraction of the self from historical location which achieves a kind of melancholic arrest in frozen contemplation. To invoke the psychoanalytic paradigm of bipolar illness, this arrest constitutes the depressive trough which is the counterpart and consequence of the energetic restlessness of melancholy's "manic" phase. The shared reverie of solitary author and solitary reader lost in the abstruse fog of the *Anatomy*'s mirages and fragments functions as a prose analogue to the dramatic ad hoc community of perception which I have argued Shakespeare made possible through the overhearing of melancholy asides in *Hamlet*. Burton's authorial legacy to his reader is to plunge them into an experience of dreamy fragmentation which has all the characteristics that Benjamin described as definitive of the allegorical representation of melancholy which he detected in German baroque literature: "the secret, privileged knowledge, the arbitrary rule in a realm of dead objects, the supposed infinity of a world without hope."[27] But how does he produce this antihistorical experience of the melancholic archive? What rhythms and tactics make this variable distance between author and source possible? If the melancholic tradition is already understood as a textual problem space defined by the tension between Galen and Aristotle, how do Burton's tactics of assemblage resolve or reformat that problem space?

Assemblage as Argument: Burton on the "Inward Causes" of Melancholy

To answer these questions, one must arrest the flowing river of Burton's thought and analyze its contents, but the entirety and integrity of the *Anatomy* forces us to make a selective cut into its plane of immanence in search of points of revealing intensity, and to risk violating its formal qualities in the process. To execute a close reading of a representative "core sample" of Burton's text is an inherently risky endeavor, for such a critical enterprise begs a number of questions: Given the vast labyrinth of his writing and the broad discursive range of generic modes within it (sermon, satire, receipt, list, digression, rant, medical text), in what sense can any particular excerpt be said to be representative? Furthermore, in what sense can one even perform a 'close reading' of a text that is composed of loose translations and excerpts of other texts? Insofar as Burton's anatomy is built out of the mangled limbs and flesh of other author's works, there is no authorial "there" there to dissect. Building up the expectation that at some point we must touch stable authorial bedrock, a typical sentence of Burton's text will lead his readers across anecdotal narratives, promising theoretical hypotheses and potentially straightforward assertions, only to terminate in a coy admission that all along he has only been passing on someone else's claims, and not speaking for himself. In large part, this problem is what has tended to drive critics in search of portable Burtonian exemplary material toward the preface, though even there the complex question of the distance between Democritus Jr. and Burton remains. But in so doing, they have avoided most of Burton's work and, furthermore, have neglected to attend to the synthetic ambitions of Burton's engagements with his sources "on the ground."

I have not, of course, selected a text at random, but rather one which happens to turn upon an exemplary shift in Burton's thinking as it passes from a Galenic physiological explanation of the causes of melancholy as an illness toward an Aristotelian account of the benefits and pleasures of melancholy as an experience. Building off of but also transforming this fundamental intellectual division, Burton's refusal to choose between these possible melancholic meanings within this core sample, and within the *Anatomy* as a whole, brings the dialectical conflict between Galenic physiology and Aristotelian proto-psychology to a suitably melancholic standstill. Yet this antisystematic stasis rearticulates at the level of argument the same melancholic performativity which his practice of authorial self-effacement through the relentless assemblage of citations expresses throughout the text. The refusal to take intellectual sides and the disappearance into fragments taken from the texts of others are two ways of failing to show up for the reader as a responsible, reliable discursive agent; they are artistic choices that produce interpretive desire on the part of the reader to verify or resolve a telling, flirtatiously unresolved sequence of gaps and disclosures offered by the author. In short, in the core sample that I have selected, we see Burton transposing into scholarly prose the generatively indeterminate dynamic of melancholy representation. Throughout this passage, as in *Anatomy* as a whole, the implied authorial personhood of Burton / Democritus is sidestepped in favor of a transpersonal assemblage whose texture is woven out of other found texts and borrowed materials.

The core sample in question is Partition I, Section III, Member III, Subsection I, a passage of five pages' length. Having described "Symptoms of Head-Melancholy," "Symptoms of Windy Hypochondriacal Melancholy," "Symptoms of Melancholy abounding in the whole Body," and "Symptoms of Maids', Nuns', and Widows' Melancholy," Burton then ambitiously promises to reveal the "Immediate Cause of these Precedent Symptoms." It is a jarring reunification; having discussed these separate modes of melancholy, the anatomically intuitive distinction between head, intestines, and "whole body" disease is suddenly riven by a gendered split which suggests that women's ailments are first a matter of their sex and secondly a matter of anatomy (they are women first, and humans with heads and guts second).[28] But here in this core sample, all four distinct subcategories are brought back together and considered causally as a new whole. At this point in the *Anatomy*, this passage is Burton's most developed and ambitious attempt to propose a "macro" level explanation for what causes melancholy. Given the pileup of symptoms and species of melancholy which he has just elaborated in the previous four subsections, and given that the pressure to deliver upon his promise to definitively resolve the question of what causes melancholy is only growing ever more acute as the reader nears the conclusion of the first partition as a whole, Burton's teasing use of the singular (*an* "Immediate Cause" will be revealed) promises to produce a decisive answer, a Grand Unified Theory that will dovetail a diverse swathe of cases into a single, etiological principle. Adopting his most therapeutically sincere tone, he begins with a flourish of protorationalist zeal:

> To give some satisfaction to melancholy men, that are troubled with these Symptomes, a better meanes in my Judgement cannot be taken, then to shew them the causes whence they proceed, not from Divels, as they suppose, or that they are bewitched or forsaken of God, heare or see, &c. as many of them thinke, but from naturall and inward causes, that so knowing them, they may better avoid the effects, or at least endure them with more patience.[29]

Gloomy metaphysical beliefs in God's abandonment, or the spiritual efficacy of demons, is traded for a robustly somatic account of what causes melancholy, with Burton conveniently forgetting the colorful catalogue of demonic causality which he himself offered up earlier in the first partition. The carping and mocking Democritus is here set aside, and Burton offers to pour forth instead a cooling stream of common sense so direct and demystifying that one scarcely notices that the singular "Immediate Cause" has already split into the plural "naturall and inward causes."

Burton's account of these causes offers a telling example of his approach throughout the *Anatomy*: when in doubt, add sources. He first acknowledges textual authority, develops our awareness of the disputes and controversies surrounding that authority, and then pushes off from that toward his own syncretic and tentative contributions to medical controversy. Characteristically, Burton begins his survey with Galen:

> For *Galen* imputeth all to the cold that is blacke, and thinkes that the spirits being darkned, and the substance of the Braine cloudy and darke, all the objects thereof appear terrible, and the *minde* itself, by those darke, obscure, gross fumes, ascending from black humours, is in continual darkness, feare, & sorrow, divers terrible monstrous fictions in a thousand shapes & apparitions occurre, with violent passions, by which the braine and Phantasie are troubled and eclipsed. *Fracastorius lib 2. de Intellect will have cold to be the cause of Fear and Sorrow; for such as are cold; are ill disposed to mirth, dull and heavy, by nature solitary, silent, and not for any inward daknesse (as Physitians thinke) for many melancholy men dare boldly bee, continue, and walke in the darke, and delight in it; solum frigidi timidi*: if they be hot, they are merry; and the more hot, the more furious, and void of feare, as we see in mad-men: but this reason holds not, for then no melancholy, proceeding from choler adust, should feare. *Averroes* scoffes at *Galen* for his reasons, and brings five arguments to repell them, so doth *Herc. de Saxonia: Tract. de melanch.* cap. 3. assigning other causes, which are copiously censured and confuted by *Aelianus Montaltus, cap. 5 & 6. Lod. Mercatus de Inter. morb. cur. lib. 1. cap. 17. Altomarus. cap. 7. de mel. Guianerius tract 15. cap. 1. Bright cap. 17. Laurentius cap 5. Valesius med. contr. lib. 5. cont. 1. Distemperature* they conclude, *makes blacke juice, blacknesse obscures the spirits, the spirits obscured, cause feare and sorrow. Laurentius cap. 13.* supposeth these black fumes offend especially the *Diaphragma* or Midriffe, and so *per consequens* the mind, which is obscured as the Sun by a cloud. To this opinion of *Galen*, almost all the *Greekes* and *Arabes* subscribe, the *Latines*

> new and old *internae tenebrae offuscant animum, ut externae nocent pueris*, as children are affrighted in the darke, so are melancholy men at all times, as having the inward cause with them & still carrying it about. Which blacke vapors, whether they proceed from the blacke blood about the heart, as *T. W. Jes.* thinkes in his Treatise of the passions of the minde, or stomacke, splene, midriffe, or al the misaffected parts together, it boots not, they keep the minde in a perpetuall dungeon, and oppresse it with continuall feares, anxieties, sorrowes, &c.[30]

At first glance, this passage exemplifies the forbidding density of half-digested sources that has permanently hobbled Burton's accessibility to post-seventeenth-century readers. Of course, part of this difficulty is simply a matter of format; imagine the results if the numerous footnotes beginning with "see also . . ." that proceed to list the available literature were simply inserted into the middle of the sentences they typically append in modern academic prose. If the prime difficulty in reading the *Anatomy* is simply the sheer volume of exemplary material it contains, Burton's tendency to hopscotch around the texts he quotes (now supporting, now refuting, now quoting someone else refuting the refutation), when combined with his habit of suddenly vanishing with an abrupt "&c," makes the thread of his argument perilously slender. Surveying the survey above, it is far from clear where exactly Burton is headed within this sequence of switchbacks across centuries of medical dispute.

We begin with Galen asserting that the darkness of melancholic humors clouds the mind, articulating a causal link between humors in the body and emotions in the mind/brain through a metaphor of relative opacity or transparency, with light and warmth implicitly aligned against dark and cold. We then shift to Fracastorius' counterclaim, against Galen, that this causal link is not based in any way upon light or darkness but entirely upon warmth or cold, with Burton reeling off an extensive quotation from Fracastorius to that effect. Cutting left, Burton then refutes Fracastorius' refutation with a "but this reason holds not," and provides a counterexample by invoking the heat-based causal theory of melancholy adust in which black bile is burnt by the heat of the Liver and converts to a new substance (confusingly both the substance and the illness are called "melancholy adust"), which circulates in the body and triggers the mental and physical symptoms of melancholy illness. Already, Burton has put forward Galen's view, then put forward a critique of it, and then, in the process of critiquing the critique, has introduced a new component which seems to uphold Galen until we notice that now, in the process of that introduction, it is heat, and not "light vs. darkness," which is emerging as the new front-runner in this discussion of the primary cause of melancholy. That is, Burton's seeming defense of Galen against Fracastorius winds up obscuring both the original territory they shared (for note that Burton at first said "Galen imputeth all to *the cold* that is blacke"; italics mine) *and* the terms of their difference. This torturous back and forth continues. Averroes and Hercules of Saxonia are brought in to attack Galen, and then a laundry list of Galenic defenders is produced in turn.

Galen, "the common Master of them all, from whose fountain they all fetch water," was clearly an important authority for Burton, and he is the most frequently cited author within the *Anatomy*.[31] Yet, despite the long shadow of his overarching influence and Burton's deep familiarity with his acolytes and followers, Galen's embattled authority is acknowledged only to be passed over as Burton's attention wanders elsewhere. Despite his reputation in some of the critical literature, Burton does not exactly cower before classical precedent; rather, he seems to delight in smothering one author with another. Galen's darkness metaphor does not return until Laurentius is mentioned, and his metaphor of the cloud obscuring the sun brings us back to where we began, with black bile imagined as a kind of psychic tint that darkens inner temperament and outer perception. In a final maddening twist, Burton's reference to the "T. W. Jes." (Thomas Wright) text *The Passions of the Minde* gives the game away by betraying a deflating lack of interest in determining the origin of black bile itself; it may proceed from the blood around the heart, the spleen, or the midriffe, but, really, "it boots not." Here it seems that Burton's own curiosity at determining the precise source of the "inward and natural" cause, the ostensible topic of the subsection as a whole, simply gutters out.

Pulling back from the fracas of claims and counterclaims, Burton offers a surprisingly tender defense of the one party who occupies the center of this discussion but whose own perspective seems oddly missing from the *Anatomy*: the melancholic himself.[32] Burton's generic frame shifts abruptly from the diagnostic treatise, and he takes on a tone of righteous moral defense that, at least at first, approximates a sermon:

> It is an ordinary thing for such as are sound, to laugh at this dejected pusillanimity, & those other symptomes of melancholy, to make themselves merry with them, & to wonder at such, as toyes and trifles, which may bee resisted and withstood, if they will themselves: but let him that so wonders, consider with himselfe, that if a man should tell him on a sudden, some of his especiall friends were dead, could he choose but grieve: or set him upon a steepe rocke, where he should be in danger to be precipitated, could hee be secure? his heart would tremble for feare, and his head be giddy. *P. Bayrus Tract de Pest. gives instance* (as I have said) *and put case* (saith hee) *in one that walkes upon a planke, if it lye on the ground, he can safely doe it: but if the same planke be laid over some deepe water, in steed of a bridge, hee is vehemently moved, and 'tis nothing but his imagination, forma cadendi impressa, to which his other members and faculties obey.* Yea, but you inferre, that such men have a just cause to feare, a true object of feare, so have melancholy men an inward cause, a perpetuall fume and darknesse, causing feare, griefe, suspicion, which they carry with them, an object which cannot be removed; but stickes as close, and is as inseparable as a shadow to a body, & who can expell, or over-run his shadow? remove heat of the Liver, a cold stomack, weak spleene: remove those adust humorous and vapours arising from them, blacke bloud of the heart, all outward perturbations, take away the cause, and then bid them not grieve nor feare, or be heavy, dull, lumpish, otherwise counsell can doe little good; you may as well bid him that is sicke of an ague, not to be adry; or him that is wounded, not to feel paine.[33]

Burton here attempts to locate melancholy in a causal account which straddles two distinct subject positions: what is laughable delusion when it is converted into a diagnostic case history is inescapable and terrifying when it shows up as firsthand experience. How can he do justice to both perspectives? And if melancholy is an inner perception or experience, what is it a perception of? In his critical appreciation *The Psychiatry of Robert Burton*, Bergen Evans eagerly produces this moment as a sign of Burton's humanitarian foresight, his prescient capacity for empathy in an age that exhibited the mentally ill as figures of fun.[34] But it also captures in a uniquely vivid fashion Burton struggling to conceptualize melancholy's hybridity, its puzzling capacity to be both as thingly as a black liquid juice ("an object which cannot be removed") and as evanescent as a moment's fleeting thought.[35] Burton has lurched from a physiological discussion of inward cause to a psychological discussion of inner experience, but coming so far in advance of the explicit articulation of an empiricist philosophy, he seems to shift nervously around the question of the ontological status of the "object" which is the cause of melancholy. From "inside" the perceptions of the melancholic would seem to partake of the robust immanence of everyday seeing and hearing of cliffs and planks, and yet the diagnostic "outside" from which this defense is issued simultaneously demonstrates that some things which seem inescapably real are in fact shadowy "phantasie." Burton's text manifests an empathetic, even democratizing, impulse: The melancholic cannot choose not to perceive the melancholy object before them, just as the healthy person could not choose not to feel sorrow if told of a death or fear if positioned on a cliff. But this impulse is also one which casts doubts on the access of therapy. Insofar as one cannot choose but feel them, what hope of relief can there be from inescapable perceptions? Once posed in this manner, the reader is reminded of Burton's opening exhortation that, if the knowledge of an inward cause did not itself remove the symptoms, it could at least produce a new stance of patient understanding toward them.

Burton's imaginative capacity to inhabit the minds and emotions of melancholics without passing judgment is arguably critical to his text's therapeutic success. But it also tends to trigger a somewhat less sympathetic repetitive gesture within his prose: the colorful case history of madness. As the core sample progresses, Burton's tender evocation of the inescapability and seeming self-evidence of the melancholy symptom when experienced "from inside" is followed close behind by a series of exotically fanciful accounts of "mad" beliefs which belong to the modern clinical terrain of paranoid psychosis and schizophrenia, and his comic relish in retelling symptomatological tales seems to rather undo the kindly subjective defense of melancholy logic which he has just advanced:

> Suspition followes Feare and Sorrow at heeles, arising out of the same fountaine, so thinks Fracastorius, that Feare is the cause of Suspition, and still they suspect some trechery, or some secret machination to be framed against against them, and stil they distrust. Restlesnesse proceeds from the same spring, variety of fumes makes them like and dislike. Solitarinesse, avoiding of light, that they are weary of their lives, hate the world,

> arise from the same causes, for their spirits and humours are opposite to light, feare makes them avoid company, and absent themselves, least they should be misused, hissed at, or overshoote themselves, which still they suspect. They are prone to Venery, by reasons of winde. Angry, waspish, and fretting still, out of abundance of choler, which causeth fearful dreames, and violent perturbations to them, both sleeping and waking: That they suppose they have no heads, flye, sinke, they are pots, glasses, &c. is wind in their heads. Herc. de Saxonia doth ascribe this to the severall motions the animall spirits, their dilation, contraction, confusion, alteration, tenebrosity, hot or cold distemperature, excluding all materiall humours. Fracastorius accounts it a thing worthy of inquisition, why they should entertaine such false conceipts as that they have hornes, great noses, that they are Birds, Beasts, &c. Why they should thinke themselves Kings, Lords, Cardinals. For the first, Fracastorius gives two reasons: One is the disposition of the body: the other, the occasion of the phantasie, as if their eyes be purblind, their eares sing, by reason of some cold, and rume, &c. To the second, Laurentius answeres, the Imagination inwardly or outwardly moved, represents to the understanding, not inticements only; to favour the passion, or dislike, but a very intensive pleasure followes the passion, or displeasure, and the will and reason are captivated by delighting in it.[36]

Burton's list of fanciful beliefs belongs squarely in the broad tradition of psychological manuals, with more or less precisely the same tales of delusional melancholics appearing in Bright, Wright, du Laurens, and Descartes, to name only four examples.[37] Despite Burton's rejection of occult or demonic etiology in this subsection, the subterranean proximity of madness and magic is here signaled in the oddly analogous transformations of the body which these melancholic symptoms share with early modern magical spells and receipts, a similarity one senses in, for example, the following spells from the commonplace book of Richard Hill (1535) which promise: "To make each man seem to have two heads, to make a man seem to have two horses' heads, that all utensils in the house may appear to be serpents, [. . .] that men may appeare to be on fire."[38] When placed in such a discursive context, the effects of madness seem so closely akin to the effects of possession and demonic influence (suddenly overtaking the victim with distorting falsehoods) that Burton's equivocal stance on their causal overlap begins to appear to us as an early modern form of empiricist "common sense" rather than the realpolitik hedging of a science writer toiling in the superstitious wake of James I's demonology. The resemblance between the hidden effects of melancholy causality upon the imagination and the "actions at a distance" affected by witchcraft is a recurring theme in the *Anatomy*; in his final partition's analysis of religious melancholy, Burton will come to speak of melancholy as a "*Balneum Diaboli*, the Divells bath," figuring black bile for the reader as a kind of witches' brew of associative and imaginative flux endowed with a shape-shifting power of imaginative transformation.[39] The material qualities of melancholy as a dark liquid humor lend a certain intuitive rightness to this cluster of associations with the poetics of sympathetic magic (not to mention the magical and assemblage-making qualities of poetry itself as an activity of combinatorial "making").

If the shift from "cause" to "causes" constituted the first decisive conceptual jump within the core sample, the paragraph above contains the second and third, in quick succession. They are easy to miss, arriving as they do within paraphrased quotations which come on the heels of these picturesque accounts of delusion: First, Burton's quotation of Fracastorius' explanation for the plural sources of these comically extreme symptoms neatly identifies the tectonic fault line that runs across the conceptual history of melancholy: the relationship between the psychological and the physical (a holistic feedback loop of interrelation doomed to dualist divorce). Second, if the symptoms associated with bodily melancholy show up as pain and discomfort, the experience of melancholy as a mental event introduces another conceptual division: melancholy pain and melancholy pleasure, here placed in a temporal sequence with pain leading to pleasure.

This acknowledgment of the "captivating" pleasure that the workings of a melancholic imagination can produce within the mind of the solitary melancholic makes good upon the promise of the poetic "Author's Abstract of Melancholy" which follows the frontispiece and precedes the *Anatomy*'s preface:

> When to my selfe I act and smile,
> With pleasing thoughts the time beguile,
> By a brooke side or wood so greene,
> Unheard, unsought for, or unseene,

> A thousand pleasures doe me blesse,
> And crowne my soule with happinesse.
> All my joys beside are folly,
> None so sweet as melancholy.[40]

Hailing melancholy as a bucolic retreat from others, this verse celebrates solitary privacy as a precursor to a kind of liquid, erotic dissolution on the hazy border between monastic contemplation and masturbatory furtiveness. Melancholy is framed here as an unproductive and antisocial pleasure, a narcissistic circuit of self-regard figured by the "smile" and "act" delivered to the inviolate self. The landscape borrows from the scenic repertoire of pastoral seduction, but in this melancholy revision the beloved and the shepherd are the same person. This passage expresses a curiously circular linkage between environment and mind, and defines the pleasure of melancholy as a kind of perfect fit between external isolation and internal calm—but Burton's quote from Laurentius reminds us that this bliss can only occur when "will and reason" have been surrendered. As with the paralyzing fears alluded to earlier, it is not chosen, but felt. While some of its readers were no doubt propelled onward through Burton's prose by their own psychological distress, his admission by way of Laurentius that it is a pleasure to view things through a lens of gentle melancholy also acknowledges what propelled even more of his readers through his text: The clearest evidence for the pleasure of melancholic contemplation is the widespread, enduring popularity of the *Anatomy* itself. Of course, to claim that it is pleasurable to read the *Anatomy* is not necessarily to claim that it is a specifically melancholy pleasure to do so. But the family resemblance between the flowing river of Burton's prose and the picturesque "brook side" of genial melancholy conjured by his "Author's Abstract" suggestively elides both acts of solitary travel across a landscape into one.

Laurentius' remark about the pleasure attendant upon the movement of the passions allows Burton to then turn toward "the Philosophers of Coimbra,"[41] a reference to the Moorish schools of medicine at Conembrica, in Portugal:

> Why Students and Lovers are so often Melancholy, and mad, the Philosophers of Coimbra assigne this reason, because by a vehement & continuall meditation of that, wherewith they are affected, they fetch up the spirits into the Braine, and with the heat brought with them, they incend it

> beyond measure: and the cells of the inner senses, dissolve their temperature, which being dissolved, they cannot performe their offices, as they ought.[42]

Here the causality of melancholy is quite glaringly reversed; in the earlier sections of the core sample, it was a bodily imbalance or overheating or excess at the level of the physically material humor itself that then produced a mental symptom or outcome. But for this group of lovers and students, it is their conscious decision to meditate "vehemently" upon a subject (either an academic problem or the image of the beloved) that then triggers a physical result: Spirits are called up to the brain and "incended," and the outcome of this excessive mental operation is a physiological disabling in which, confusingly, "the cells of the inner senses" (a woolly compromise between somatic and psychological registers) are said to be "dissolved." The shift from a depiction of excessive heat as a primary cause of melancholy to a depiction of excessive heat as a consequence of excessive meditation prepares the way for genial melancholy by relocating causal responsibility in the mind; in addition, this connection between students and lovers in the context of a preceding description of the pleasures of melancholy passions foreshadows the introduction of genial melancholy by associating melancholy with both pleasurable experience and intellectual achievement. Following from Laurentius' reference to the pleasures of melancholy, the digression on the philosophers of Coimbra, in turn, produces the first direct appearance of the thirty-third pseudo-Aristotelian problem into Burton's text, a full four hundred and twenty-one pages after it has begun:

> Why Melancholy men are witty, which *Aristotle*, hath long since maintained in his Problems: and that all learned men, famous Philosophers, and Law-givers, *ad unum fere omnes Melancholici*, have still beene Melancholy; is a problem much controverted. *Jason Pratensis* will have it understood of naturall melancholy, which opinion *Melancthon* inclines to, in his booke *de Anima*, and *Marcilius Ficinus de san. tuend. lib. 1. cap. 5.* but not simple, for that makes men stupid, heavy, dull, being cold and dry, fearfull, fooles, and solitary, but mixt with the other humours, fleagme only excepted: & they not adust, but so mixt, as that blood be halfe, with little or no adustion, that they be neither too hot nor too cold. *Apponensis* cited by *Melancthon*, thinkes it proceeds from melancholy adust, excluding

> all natural melancholy, as too cold. *Laurentius* condemnes his *Tenent*, because adustion of humours makes men mad, as Lime burnes, when water is cast on it. It must bee mixt with blood, and somewhat adust, and so that old Aphorisme of *Aristotle* may bee verified, *Nullum magnum ingenium sine mixtura dementia*, no excellent wit without a mixture of madnesse.[43]

This is something more and something less than an endorsement of the pseudo-Aristotelian position. Burton's citational remove and zeal for fine distinctions threatens to whittle the thread of his argument down to invisibility. The phrasing of the first sentence grants the claim that melancholy men are witty, handling it lightly as a *fait accompli*. The task, it seems, is to explain *why* it is true, not to question *whether* it is true—and yet the conclusion of the sentence in the phrase "is a problem much controverted" already raises the specter of the directly opposite conclusion: that the pseudo-Aristotelian Problem itself *has* been contradicted. Which, of course, it has been. Being cold and dry according to traditional humoral principles, "simple" melancholy (i.e., the pure black bile within the body), which has been neither burnt to an adust nor mixed with other humors, produces a heaviness and dullness which is precisely at odds with the Aristotelian claim; the outcome of a surfeit of this cold and dry humor would be a sluggish inability to speak and perform even the most simple tasks, and would of necessity smother any preternatural capacity to dazzle and excel in art, literature, or politics. Quick to enlist support, Burton dutifully produces texts that explain this discrepancy between the Aristotelian model of genial melancholy and the normatively humoral Galenic/Hippocratic model. Jason Pratensis, Melancthon, and Marcilis Ficinus (Ficino) are impressed into service to provide support—but the solution itself is articulated not through a "cultural" response (i.e., a list of distinguished "genial" melancholics cross-indexed with their achievements in art, science, or politics) but through an entirely physiological explanation which seems to validate the Aristotelian claim, but in terms set by the Galenic/Hippocratic model.

"Simple" black bile, when mixed with other humors (but not phlegm, and not burnt into an adust), can, it seems, trigger that celebrated symptomatic creature, the witty melancholic—or so Burton's strangled half sentence at first seems to imply. No sooner has this recipe for witty melancholy been provided then it is instantly controverted by Apponensis via Melancthon, who specifically insists upon the causal role of melancholy adust. But, in a classic Burtonian triple play, Laurentius is now brought in to controvert the controverter, and the causal responsibility of melancholy adust is now itself, in turn, called into question. But does Burton believe that two wrongs make a right? If Ficino is refuted by Apponensis, and Apponensis is refuted by Laurentius, does that mean that Ficino has been verified? Yes and no, for Burton's digressive tendencies suggest that his controversial to-ings and fro-ings do not so much define a pendulum swing between binary oppositions so much as a corkscrew motion in which, on the way to an imagined return to a first position or topic at hand, his own mind has already changed, and so each "return" is a transformative relocation of its stakes and its sense.[44] Boiling down the admittedly murky prose, according to Burton, Ficino is right for having insisted upon a mixture but wrong for refusing to allow for adust within that mixture.

Having brought his triple play to a close, Burton now provides a new recipe for the production of a mixture through which Aristotle's claim "may be verified," but it is a recipe soured by Burton's vague phrasing: It is not clear whether he is saying that 1) simple melancholy, when mixed with blood that is somewhat adust, can produce the outcome of a witty melancholy, a genius mixed with madness, or whether he is saying that 2) simple melancholy that is somewhat adust, when mixed with blood, can produce the outcome of a witty melancholy, a genius mixed with madness. Bracketing the specific question of which humor may be adust, his overall claim is that the mixture of these affective traits must be a direct and literal expression of a somatic mixture of humors. A relation of correspondence must obtain between inner cause and outer result, and yet Burton seems here to be reading backward from effects to causes by way of a hasty compromise between positions which, if considered closely, are incommensurable. The Galenic belief that black bile is cold and dry and the pseudo-Aristotelian belief that melancholics are witty are accommodated through a humoral sleight of hand in which internal cookings and mixtures are invoked in order to splint the fracture between humoral physiology and pseudo-Aristotelian psychology. If Burton's layman respect for Galenic medical authority leads him to patch up a solution in biological terms, Burton's horse sense about human beings leads him to do so in order to support Aristotle.[45] Sounding the securely proverbial ring of "no excellent wit without a mixture of madnesse," Burton ends our "core sample" by moving on again, restlessly, to partitions new.

Since at least the era of the Warburg Institute and Lawrence Babb's *The Elizabethan Malady*, the binary opposition between the Galenic and the Aristotelian models of melancholy has been treated as a more or less stable macrolevel position within early modern intellectual history which contours the available states of melancholic meaning at any given point in the period. From the vantage of this standard account, one can argue that Burton fails to decisively resolve the causality of melancholy because there is no neutral vocabulary available to him, no set of terms or models for the organic relationship between embodiment and psychic life which do not symptomatically exhibit the conceptual cleft at the core of melancholy's fractured intellectual history. Alternately, we can adopt the skeptical position of Fish and read Burton's "failure" to resolve the intellectual problems he sets himself as a sarcastically rigged excursion down a blind alley, and nothing more. But perhaps such expressions of disappointment or condescension are themselves premature and miss the extent to which Burton's equivocal engagement with melancholy causality is an exemplary moment in which, far from failing to swing polemically toward one side or the other of a retroactively defined impasse, he is succeeding in thinking (and writing) melancholy dialectically.

The Burtonian textual construction of melancholy as endlessly belated, equivocal, and overdetermined presents it to the reader as a node of conflict rather than an object of knowledge. Burton models melancholy as antagonism, as the site of an active struggle between opposed forces, in a manner which far more successfully expresses its real status *as an Idea* than either a decisively "Galenic" *or* "Aristotelian" position would realize. Taking in ancient Greek writers and Burton's living contemporaries and, more importantly, refusing to choose between them, the broad reach across the available literature within this core sample reveals Burton's antihistorical, panoramic methodology. If the Galenic/Aristotelian split in itself cannot be said to directly structure the *Anatomy* because Burton engages entire troops of textual authorities and builds his text out of them through macaronic quotation, the fundamental result of this split—melancholy as a condition which is defined by its overdetermined position in the midst of conflicting understandings and mutually incompatible intellectual histories—proves decisive for Burton's text, for he fashions his text out of that conflicted heritage, capturing the state of available thought across eras and cultures rather than isolating and defending a local territorial position within it. It is here that Benjamin's insight into the dialectical character of the intellectual history of melancholy usefully frames Burton's achievement in the *Anatomy*; describing Saturn's doubled influence (as an earthly, cold, and dry planet, it produces "material" individuals suited only for agricultural labor, but as the highest planet it produces sages suited to "the most extremely *spiritual religiosi contemplativi*"), Benjamin wrote that "[t]he history of the problem of melancholy unfolds within this dialectic."[46]

There is a stylistic, positional unity of perspective upon this mass of received historical material that disintegrates its linear sequence in order to make it available to the reader as a panorama. Burton is uniquely useful not as an indicator of what was emergent and what was receding but rather because his indifference to such claims is the marker of his melancholic relationship to the task of scholarly knowledge; his shuffling of the deck confirms Lauren Shohet's observation that "anachronism itself foregrounds tradition as use, rather than recovery, allowing audiences to access historicity, if not history."[47] Instead of a vertical ascent from superstition to science or from ignorant darkness into enlightenment, Burton's text imposes a horizontal drift across a fragmented landscape whose rubble, gathered from across a swathe of historical and cultural range, now lies pulverized before him and his reader.[48] It is not quite correct to say that he surveys rather than chooses—for his acts of choice about what to quote and when are consummate editorial decisions—but his artfulness emerges in the way that he seems not to choose. In depicting Burton as a surveyor, I have in mind, of course, Anthony à Wood's biographical sketch of Burton in *Athenae Oxoniensis*: "He was an exact Mathematician, a curious Calculator of Nativities, a general read Scholar, a thro'-pac'd Philologist, *and one that understood the surveying of Lands well*."[49] Requiring precise observation but also a number of potentially conflicting perspectives, surveying makes an apt metaphor for Burton's approach to the task of describing and delimiting a broad, organic, and unruly field of knowledge. Furthermore, it models the immersive character of his own relationship to the landscape he surveys; far from operating at a comfortable remove from his object of study, Burton is himself writing *as* a melancholic as he writes *upon* melancholy.

Burton's editorial stance in relation to his own text only furthers this antihistorical methodology: For Burton, as both author and editor, addition is endlessly possible, but that is not quite the same as "progress." As the *Anatomy* itself entered the discursive history of science texts along with the predecessors it incorporated, it simply took on more material without any particular effort at "improvement." Much was added, but almost nothing was corrected or removed aside from occasional printer's errors and typographic infelicities; in Lawrence Babb's memorable phrase, "it grew like a vegetable," adding seventy percent more material between Burton's first edition in 1621 and the last edition he lived to publish in 1638.[50] By the time of his death two years later, the printer who brought out the posthumous sixth edition in 1651 had even more additions to incorporate into the text, regardless of Burton's repeated assurances that he was "now resolved Never to put this Treatise out again."[51] Moreover, the new material did not consist in updates based upon particular scientific or medical innovations but was spread evenly across the historical spectrum: As it moved forward in historical time, the *Anatomy* simply kept expanding outward in size and backward in panoramic range, with Burton incorporating new quotations from Plato, Virgil, and Ovid, but also from contemporaries such as Hercules de Saxonia, Scherertzius, and Frambesarius.[52]

Burton's refusal to choose sides within the Galenic/Aristotelian antinomy, and his antihistorical practice of leveling down medical science's 'winners' and 'losers' into an inclusive catalogue that fails to draw the distinctions one expects given the ostensibly therapeutic and practical dimension of his project, models a kind of scholarly melancholia, a structural arrest within the accumulated materials of knowledge that freezes thought in place on the threshold of a decisive reaction. It produces a kind of contemplative immersion which invites the reader to participate in a second-order melancholic affect of passive flow and reverie uncoupled from the argumentative rhetorical responsibility of taking sides. If the continual foregrounding of dispute between sources models melancholy as a site of dialectical conflict, the refusal to take sides seems to halt the progression of dialectics toward its resolution, placing the author and the reader within an open-ended pause always "just before the moment of truth" in which conflict will produce an outcome. This indecisive retreat from conflict into the stasis of contemplation is itself one form of melancholy, the form that the *Anatomy* renders portable and accessible to the reader. One could also describe this frozen state of contemplative arrest in the midst of the mute fragments of an antihistorical panorama as "dialectics at a standstill." The phrase is Walter Benjamin's, and it is one that he used to characterize his own methodological ambitions, to which we now turn.

Digression as Argument: Benjamin on the "Mosaic" of Melancholy

How did Benjamin imagine the relationship between the individual citation and the work as a whole? What might Benjamin's thinking make available to readers of the *Anatomy*? To grasp the peculiar demands which Benjamin's thought places upon the part-whole relationship between text and exemplary quotation, however, one must engage in one of Burton's own signature "bad habits": digression. In particular, a digression into the critical methods which Walter Benjamin developed in his first work, the *Ursprung des Deutschen Trauerspiels* (hereafter referred to as *The Origin of German Tragic Drama*), and in particular its "Epistemo-Critical Prologue." This digression is necessary if we are to understand Benjamin's methodological philosophy. This prologue, with its antihistorical Platonist temporality and celestial metaphors of ideas as constellations and concepts as stars, famously wrong-footed the faculty of *Germanistik* at the University of Frankfurt am Mainz, and led them to forward Benjamin's *Habilitationsschrift* to the faculty of Philosophy and Aesthetics and, upon their advice, ultimately to reject the *Ursprung* in its entirety.[53] In surveying the site of this first scholarly repudiation, we should resist any too easy identification with its tragic Benjaminian critic-hero; we must place each step with the care required when breaking and entering upon any other condemned building.

The prologue begins with a call from outside itself, an opening quotation from Goethe's *Materialen zur Geschichte der Farbenlehre* that stands as a motto for Benjamin's own project throughout the *Ursprung*: "[S]cience ought to reveal itself completely in every individual object treated."[54] Functioning as the *Ursprung*'s own very first object, this Goethe quotation itself performs the function it describes, as it articulates the entirety of the methodological and critical labors to follow, commencing them, and exemplifying their spirit, with the singularity of a thunderclap. Benjamin is calling for a methodology of immanent philosophical truth, not merely a ratio of propriety between "form" and "content" but something far more radical, which is cued if we transpose Goethe's verb "to reveal" into its noun form: revelation, a word heavy with messianic import. To grasp the force of Goethe's insight within the project as a whole, we must attend not only to the quotation's literal meaning but to its affective dimension, and hear the plaint within Goethe's "ought," discerning in it the expression of a wish simultaneous with the assertion of a standard: the wish that the presentation of an individual object might also be suffused by a total revelation, a revelation of totality somehow immanent to the object. Such an absolute demand lends the succeeding text its own histrionic heroism, even independently of the text's fate at the hands of its first readers, a drama in which the Benjaminian scientist/philosopher falls under the threatening shadow of an unassuageable judgment. If each moment of critical attentiveness raises the possibility of a philosophical knowledge without limits, then the stakes of a failed encounter correspondingly threatens to wound the whole: The lapse into the mute silence of "natural history" on the part of any individual object under consideration thereby becomes an absolute loss, not merely a local fumble but a negation of the totality of revelation.[55] In tracking this methodology of immanence, trying to understand how Benjamin imagined it, and how his own text struggles to perform this work of revelation through quotation, I am also hoping to draw out a series of revealing parallels with Burton's own dialectical encounter with the melancholy archive he cited from and rebuilt.

Benjamin commences the prologue by trying to imagine how philosophical form and truth might be made coextensive, adequate to each other. Philosophy is haunted by the question of its own representational status, which is ultimately the question of the adequacy of language. There can be no recourse from this dilemma through a nonlinguistic symbolization or abstraction:

> The more clearly mathematics demonstrate that the total elimination of the problem of representation—which is boasted by every proper didactic system—is the sign of genuine knowledge, the more conclusively does it reveal its renunciation of that area of truth towards which language is directed.[56]

The way out of language is possible, but only if we would also sacrifice a truth that specifically pertains to language. Benjamin's final phrase shows him literally thinking the linguistic turn, but in a uniquely immanent manner: He describes language as itself always already turned, inclining toward its target, thrust in the direction of its particular orientation, truth. Truth is itself, in turn, spatialized as an extended area, some of which we can reach through mathematics and some which we can reach only by way of language. We have then, from the start a rich terrain already densely populated with figures (*science, objects, truth, representation, philosophy, language*) whose relationship is far from clear, and as Benjamin progressively complicates his portrait of their deadlock, one can almost feel the prickling restlessness and resistance of the Frankfurt faculty. The first real blow is then struck with Benjamin's repudiation of the systematic philosophies of the nineteenth century as incomplete evasions of the representational dimension of philosophy's proper task. Attempts to construct a system cast truth as "outside" the system, and suffer from a kind of methodological bad faith as a result, system builders are guilty of a "syncretism" which struggles to unify distinct terrains with fatal consequences.[57]

In contrast to this, Benjamin lauds the theological treatise for its anti-systematic declaration of doctrine. While a glimpse at Aquinas suggests that such writings are by no means lacking in system-building exertions, Benjamin's intentions in drawing a distinction between system and treatise is clarified and immediately expanded:

> The treatise dispenses [. . .] with the coercive method of mathematics. In the canonic form of the treatise the only element of an intention—and it is an educative rather than didactic intention—is the authoritative quotation. Its method is essentially representation. Method is a digression. Representation as digression—such is the methodological nature of the treatise.[58]

We can sense here the proximity to the *Anatomy*'s prose assemblages, for Burton puts into practice Benjamin's theory of the treatise as an essentially digressive mode of representation in both his celebrated set pieces such as the seventy-page "Digression of Air" and at each discreet point in his text in which he seems to handle each source only "in passing," sauntering toward yet another transient encounter. Benjamin's text offers a richly compressed sequence of equations and conclusions that must be parsed with care if we are to avoid a flattening collapse into an "authoritarian" reading of the treatise's peculiar effect. According to Benjamin, the treatise is not engaged in convincing us, proving by argument that we ought to accept its conclusions, or leading us slowly, step by step, from minor agreements with opening premises toward a final acceptance of a larger claim. Such a piecemeal and provisional approach would instrumentalize truth and subjugate it to the status of mere knowledge. Rather, at each point the treatise *represents* truth directly. There is then no need to proceed from one position to another along a path toward a goal, for in each moment of citation or quotation in which the truth is represented before us, we are already somehow within it, and thus every movement within the treatise is a betrayal—a falling away from—the rapt habitation of a reader within the truths it represents. Paradoxically, then, the method of the treatise as a sequence of truth-representations can only be through a necessary "fall" within the movement from one to the other, a digression. This digression might be described as a molecular "hop" akin to the Brownian motion of subatomic particles, rather than through the standard definition of digression as a *lapsus* from the linear ideal of an argumentative line of flight. Benjamin's wager upon the immanent power of the fragment comes at a substantial cost; cut off by necessity from systematic argument, his license to drift places terrible pressure upon each encounter to properly represent a truth which would not be relational and local.

Benjamin is interested in the halting rhythm of the theological treatise, its capacity for contemplative arrest in each quotation, a caesura of awed respect produced by the elevated status of each quotation as doctrine; doctrine elevates the textual example to the dignity of a self-sufficient and eternal truth that annuls momentum and cancels argumentative flow. The theological example is literally staggering. This carries with it a corresponding denial of the logic of contiguity or adjacency that would allow an argument to surface in the space between quotations. In the radical self-sufficiency of its citational quanta, doctrine is here imagined as a non-ordered series of monadic eternities, side by side and yet utterly distinct from one another. Such a vision bears, it must be said, little relation to any particular example of a theological treatise. Rather, it should be read as a kind of idealized image of the theological law as a body of statements which do not found or explain or support one another, but which all partake equally of the same universal truth-content.

Aware of the strain involved in conceptualizing such a textual model, the opening remarks on the treatise are supplemented by a secondary example of the mosaic: "Just as mosaics preserve their majesty despite their fragmentation into capricious particles, so philosophical contemplation is not lacking in momentum."[59] After the example of the treatise, the metaphor of the mosaic provides Benjamin's second model for the relation between the individual object and the truth it represents. This metaphor risks a certain failure. For when we try to map the temporality of the contemplative pause occasioned by discrete quanta within the treatise onto the pictorial logic of the mosaic, their mutual metaphoric equivalence collapses. Surely, someone regarding a mosaic as a totality is not encountering it in a staggered experience of one discrete fragment at a time; they are encountering it holistically, as an overall pattern or design. To immerse one's self in a single fragment of a mosaic would be precisely to fail to perceive the mosaic's general pattern, to stall in a bad eternity, to entrap one's self in a partial component at the expense of the whole. And yet this immersion in the minute detail is precisely what Benjamin is calling for, and regarding as the axis of comparison between treatise and mosaic:

> The value of fragments of thought is all the greater the less direct their relationship to the underlying idea, and the brilliance of the representation depends as much on this value as the brilliance of the mosaic does on the quality of the glass paste. The relationship between the minute precision of the work and the proportions of the sculptural and intellectual whole demonstrates that truth-content is only to be grasped through immersion in the most minute details of subject-matter.[60]

Submerging the reader in a seemingly neverending progression of minutiae, Burton's text risks this outcome: Memorable examples seem to strangle, or at least overpower, the didactic whole. Similarly, Benjamin's insistence risks wounding the integrity of his metaphor, for precisely insofar as we are invested in the minute detail of a mosaic, we are not regarding the majesty of its overall design. Benjamin's metaphor thus functions ironically, cruelly signaling the inflexibility of the very standard that the Goethe quotation calls for: The failure of the fragment to telegraph the whole thus occasions a kind of momentary failure of the whole. The immanent leap from the single element to the whole cannot be performed with a mosaic. It seems that some other logic is needed and, oddly enough, mathematics provides a closer exemplum of this methodology of immanence. Koch curves provide a topological example of a part/whole relationship of total transparency; discrete cuts within the curve will always display at their local level of resolution the same articulation that the curve as a whole displays at a higher level of resolution—a radical sameness obtains between the "macro" and the "micro."

A characteristic example of this structural resonance across different levels of description would be the symmetrical relationship between Burton's project as a whole and his individual sentences. Burton's structural innovation does not simply subsist at the macro level of the *Anatomy*'s partitions, members, sections, and subsections. It is also expressed at the level of the sentence through his pronounced use of long winded parataxis and comically grotesque, encyclopedically inclusive lists. Burton scholar Michael O'Connell has dryly described the architecture of a typical Burton sentence as one "built of short phrases, frequently in apposition to one another, and there is an absence of subordinate clauses."[61] Burton suppresses the object of infinitives, and often drops conjunctions. In cataloguing these distinctive features, O'Connell notes that English had no formalized grammar texts to define correct and incorrect usage until the eighteenth century, and so Burton's far looser sentence structure—in which occasionally there is no verb at all, only a long paratactical list of phrases hanging off of a subject—was not technically "incorrect" in its time.[62] Burton's lists structurally embody the anatomical conceit of the text as a whole at the level of its most basic unit: the microstructure of the sentence cites and reiterates the macrostructure of the entire treatise. Copia and plurality obtain across the panorama of the work, and a kind of mini-copia persists within any discrete detail inside it. Lists also provide a kind of comic energy, an eruption of joy in the abundant resources of language that serves Burton's therapeutic agenda. The power of the *Anatomy* to hold its readers enthralled (recall Dr. Johnson's claim that the *Anatomy* was the only book charming enough to make him get out of bed an hour early in order to return to it) lies in part in the fact that any particular sentence is itself a seemingly endless staircase of additions and extensions which spiral downward from a more or less simple starting point.[63]

Burton's descriptive exempla of ways to entertain oneself in winter provide a fitting display of the rhythmic effects of such a list structure:

> The ordinary recreations which wee have in Winter, and in most solitary times busie our minds with, are *Cardes*, *Tables* and *Dice*, *Shovelboard*,

> *Chesseplay*, the Philosopher's game, small trunkes, shuttlecocke, bailliardes, musicke, maskes, singing, dancing, Yulegames, frolikes, jests, riddles, catches, purposes, questions, and commands, merry tales of errant Knights, Kings, Queenes, Lovers, Lords, Ladies, Giants, Dwarfes, Theeves, Cheaters, Witches, Fayries, Goblins, Friers, &c. such as the old woman told of *Psyche* in *Apuleius*, *Bocace* Novells and the rest, *quarum auditione pueri delectantur, senes narratione*, which some delight to heare, some to tell; all are well pleased with. (Part II, Section II, Member IV, Subs. I, 79)

At first glance, we are simply provided with a heading, and then out tumbles a hodgepodge of materials that belong underneath that heading, recalling Deleuze's definitive claim that "in assemblages you find states of things, bodies, various combinations of bodies, hodgepodges."[64] But the effect upon the reader of Burton's sequence, with its shifting levels of detail and its telescoping into "merry tales" with a sublist of their cast of characters, has an uncanny way of turning an inventory into a parade. In the midst of listing the means of delight, Burton himself takes delight in his own jingling, rambling rattle bag. Compare this with the list that commences Benjamin's *Konvolut* (categorical subsection) on the Panorama:

> There were panoramas, dioramas, cosmoramas, diaphanoramas, navaloramas, pleoramas (pleo, "I sail," "I go by water"), fantoscope[s], fantasma-parastases, phantasmogorical and fantasmaparastatic *expériences*, pictureseque journeys in a room, georamas; optical picturesques, cinéoramas, phanoramas, stereoramas, cycloramas, *panorama dramatique*.[65]

The same arc of comic energy builds up, like a static charge, as we move along this increasingly ridiculous sequence of variations. The eye checks for repetition but each unit of the list is, in fact, new, and yet the sequence of slight variations in forms of panoramic entertainment also conveys the desperate capitalist struggle for a slightly new commodified experience to market to the consumer. The list is a "pure" list, and yet there is something hauntingly argumentative about its mass, something both ridiculous and accusatory. Piling one entertainment on top of another, Burton's list provides a kind of miniature holiday, while Benjamin's list similarly indulges in a kind of paroxysm of manic overstimulation. These authors are seemingly headed in opposite directions, for Burton is stressing the therapeutic potential of such a list to remedy melancholy, while Benjamin is pointing toward the "skull beneath the skin" of capitalist alienation within the entertainment on offer in the marketplace.

Burton is swerving from melancholy and Benjamin is homing in on it; but at the level of rhythm and structure, these assemblages are united in their accumulative energy, and in the constitutive effect of that energy's proximity to melancholy as either founding obstacle or encrypted truth of the resulting energetic list of entertainments. The resulting melancholy assemblages which these authors construct are marked by a simultaneity of affective cross-currents: The slump downward and inward of despair holds our focus upon and within the isolated fragment, module, unit, sample, or quotation in a kind of corpse pose of attentive eternity, while the manic pulse of forward-hopping motion from fragment to fragment exemplifies the opposite current of an appetitive, kinetic process of ever-increasing accumulation and conjunction that threatens to never stop flowing and growing.

It is this immanence of the whole in the single detail that Benjamin claims is definitive for the appearance of the quotation within the treatise. Nor is it a pious encounter with spoon-fed rations of received doctrine which he has in mind; as he put it in *One Way Street*: "The quotations in my works are like robbers lying in ambush on the highway to attack the passerby with weapons drawn and rob him of his conviction."[66] Striking like lightning, this is a sudden interruption of the text with a voice that comes from outside the boundaries of the text, as it were, passing through it on its path toward the reader. That said, Benjamin's ideal and the practical experience of his texts' own examples are not always in perfect sync. Not every quote can hit like lightning. My goal is not to deconstruct Benjamin's project, skeptically reducing it to an illogical shambles; but there is a curious and immediate material tension between Benjamin's model of the suspended doctrinal truths of the treatise and his image of the holistic mosaic of fragments as models for the temporal relationship between text and truth. To risk belaboring his examples with an overly literal pressure, there is a level at which the two can, of course, be compared. What the quotation of doctrine is in the treatise, the broken piece of glass is in the mosaic: They are both fragments, both refugees from a prior unity, and they can both become the occasion for a contemplative immersion, which Benjamin describes as "a continual pausing for breath, [which] is the mode most proper to the process of contemplation."[67] But the stakes of this immersion, its capacity to produce revelation, seem quite different: To suspend one's reading in the doctrinal quotation within a treatise is to inhabit a truth and, for Benjamin, truth is eternal, radically outside of time.

Benjamin's overarching interest at this point in the prologue is to determine a form adequate to philosophy as an interface between representation and truth. The treatise, as a linear arrangement of citations of doctrinal truths along a particular path, both enables the possibility of our entrance into their revelatory eternity and sorrowfully betrays that possibility; in its very form, the textuality of the treatise as a prose form betrays its own doctrinal confidence and expresses a contrary hunch that we will not linger in any quotation quite long enough, that we will press on, still searching. In digressing from the exemplary quotation of doctrine within the treatise, we are admitting a certain restlessness, a certain twitch of affect, a compulsion, which betrays revelation and keeps it from happening.

This defines the dialectical temporality of melancholy as a mode of reading: a staggered flitting between experiences of losing oneself in the contemplation of the object and a restless jump onward to the next such experience of self-loss. The always-to-come, and thus "impossible," temporality of revelation makes a certain practice of hovering within the doctrinal exemplum of the treatise necessary as a liminal approach. One would expect that, in turning to the mosaic, Benjamin would assert that, by contrast, we must avert or deliberately blur our gaze from the apprehension of any particular fragment in order to grasp the "underlying idea," the representational "big picture" toward which each fragment contributes. But, as we have seen, that is precisely what he does not do. Why? The answer lies in the philosophy of the idea that Benjamin is advocating in the prologue and exploring in the *Ursprung*, and in this context we can see the fitness of the mosaic example. The very absence of temporality in the mosaic, the fact that it is spatially constructed, is precisely what makes the mosaic such a useful model for a structural yet antisystematic methodology. The relationship between fragmented part and "underlying idea" within the mosaic, in which this underlying idea is nothing more than the total arrangement of its constituent fragments, and yet the totality remains legible as an additional level of structuring significance, drives his famous dictum: "Ideas are to concepts as constellations are to stars."

What does this mean? The beauty and jewel-like compactness of this assertion has inspired more admiration than analysis, and its form has upstaged its sense. As with the mosaic metaphor, a brilliant surface is asserted to contain an immanent structural meaning, at once independent from, and utterly suffused within, its expressive elements. A brilliant surface of articulation conceals a certain counterintuitive twist. This problem occurs when we try to think about this claim temporally or historically and apply what we know about stars and constellations to Benjamin's definitions of ideas and concepts in play within the prologue. The stars predate humanity, and the human fictions that distinct societies project upon them; the organizing narratives and figurative shapes of the constellation are a textbook case of "cultural construction" at work. Furthermore, the clustering or grouping of stars within a constellation is a marker not of their actual spatial orientation relative to one another but of their appearance to us from a great distance as a resolved and flattened picture plane. The constellation then is precisely not what is "pre-given," "preexistent," independent of conscious observation, eternal, or timeless. And yet Benjamin asserts precisely this.

At this point, it is helpful to return to the solid ground of empirical objects and observations in which this seemingly most abstruse formulation finds soil. If we expand our scope to take in the initial articulation of the constellation metaphor, we see that the structurally preexistent "Idea" only shows up to us through its mediation in the concrete and ready to hand:

> The representation of Ideas takes place through the medium of empirical reality. For Ideas are not represented in themselves, but solely and exclusively as an arrangement of real, concrete elements in the concepts. And indeed as the configuration of these elements. [. . .] The staff of concepts which serves as the representation of an Idea realizes it as such a configuration [. . .] Ideas are to things as constellations are to stars. This means, in the first place, that they are neither their concepts nor their laws.[68]

How can we understand Benjamin's relation between the idea and its constituent "things"? One useful comparative model might be Althusser's notion of expressive causality—in which a structure is said to be the "cause" of its effects, and its effects are taken to be the total body of its component parts—but that structure is not separable from the contingent arrangement which expresses it: "[T]he structure, which is merely a specific combination of its peculiar elements, is nothing outside of its effects."[69] I would like to suggest that melancholy is itself such an "idea," that is, such a constellation or configuration in the specific sense that Benjamin develops it in *On the Origin of German Tragic Drama*: Melancholy does not have a propositional character, that is, it does not yield a single, united, literal propositional "meaning," but it is recognizable, it communicates, it generates an indefinite, potentially infinite series of representations which structurally express its idea, the idea of melancholy. In this Benjaminian sense, melancholy is a kind of structural complex that can generate a potentially endless series of examples. These particular examples can make acts of recognition, seeing, reading, or experiencing arrangement X or constellation Y *as melancholic* possible, whether they are artistic portraits, dramatic entanglements between literary characters in a play, living individuals in a consulting chamber, natural landscapes, or arrangements of fragments of received knowledge in prose form. The relational networks thus created are both "aspectual" and "epistemological," and yet they are not merely or simply subjective. Burton's and Benjamin's melancholy examples presuppose a viewer or reader for whom things could show up as melancholic; but in their sidestepping of an expressive moment of transmissible selfhood in favor of an ongoing, relentless, and open-ended structure of assemblages made out of anything in the world, Burton and Benjamin's discrete cases and examples also move past or beyond the particular subjective perspectival location for whom they are framed, and induce an immersion in a desubjective encounter with "material" in an expanded sense: a materiality that is both physical/thingly and discursive. Assemblages thus bind selves, physical objects, texts, and discourses together, in the process modeling melancholy as both affect and activity.

Such incomplete, melancholic open-endedness can become a way of life. But it can also become a way out of life. Burton spent nearly twenty years of his remaining lifespan adding and increasing his store of exemplary materials to the melancholic structure he first proposed in 1621: But the structural "idea" he limned, the skeleton beneath the *Anatomy*, remained stubbornly in place across all of these additions and expansions. After this long Benjaminian digression, we are now in a position to build upon the assertion that began this chapter, the assertion that Burton's text is both a dependent *cento* and methodologically "original": Burton's text expresses its structure through the total set of its constituent parts, all of which are derivative of others. Yet in his concrete arrangement of these parts as found material, Burton achieves a dialectical synthesis of their mutual antagonisms which structurally expresses at each moment in the text a core "idea" of melancholy which subtends them all but which cannot be articulated outside of that structure. This explains why the *Anatomy* is both saturated with heterogeneous and localized quanta of meaning while remaining worryingly "all the same"—Burton's immanent structure goes to work upon us as soon as we start to read and increases in power the deeper we descend (or ascend) across its terrain. Fittingly, on the farther shore of its labyrinth of connections, quotations, and evasions, Burton's assemblage leaves us with a homely final pronouncement that quietly insists upon the saving power of active multiplicity and willful relation: "Be not solitary, be not idle."

6. *My Self, My Sepulcher*

"Humours Black": Samson as Melancholic

Concluding our sequence of portraits in black, it must be said that Samson makes an unlikely candidate for melancholy. If we take up Robert Burton's portable definition of this condition as "feare, and sadnesse without any apparant occasion," then the application of this modish misfortune of early modernity to the Biblical hero from the Book of Judges seems, literally, woefully inadequate.[1] However densely overdetermined, Samson's causes for fear and sorrow are all too obvious: his compromising relationship with Dalilah; his disastrously credulous confession of the secret of his strength; the subsequent shaving of his hair and putting out of his eyes; his capture and enslavement at the hands of the Philistines; and a grim future of continued political imprisonment, public humiliation, and forced labor. One could go on. And yet Samson's affects and actions, at least as they are represented within the seventeenth-century poetic and dramatic framework of John Milton's *Samson Agonistes* (1671), seem to repeatedly produce explanation in terms of melancholic disease. Samson shows up for others as melancholic. The way he sits, the way that he moves his body, the way that he speaks, the kinds of desires he voices, the overall affective impact of his person registers for both strangers and intimates in terms that would have been familiar to Milton's readers as citations from the cultural repertoire of black bile. Samson's physical posture is instantly recognized by the chorus as such, and relayed to the readers of this closet drama in a manner that recalls atrabilious iconography:

> CHOR: See how he lies at random, carelessly diffused,
> With languished head unpropped,

> As one past hope, abandoned,
> And by himself given over; (118–121)

These are the downcast eyes of despair, of retreat from others and from outward-directed interest, that can be found in countless images from Cranach, Feti, and Dürer on the continent, and in the miniature melancholic portraits of Hilliard and Oliver executed for clients of taste in Tudor England already discussed in the first chapter of this book. In presenting Samson as "unpropped," Milton one-ups the by-now-overfamiliar melancholy propping posture by suggesting that his hero lacks even the energy for a minimal act of self-support. Samson describes symptoms that would have been familiar to patients and doctors in the period ("I feel my genial spirits droop / My hopes all flat, nature within me seems / In all her functions weary of herself" 594–596). Samson's interlocutors, particularly his father Manoa, explicitly diagnose him as suffering from an internal imbalance or *dyscrasia:*

> MANOA: Believe not these suggestions, which proceed
> From anguish of the mind and humours black,
> That mingle with thy fancy. (599–601)

What are the cumulative effects of this complex of symptomatic complaints from "inside" and humoral diagnoses from "outside" the black box of Samson's blind interiority? Are these just momentary excursions into the fluid-soaked language of humoralism, mere detours into the somatic en route to the more respectable, not to mention more modern, proto-psychological registers of "inwardness" and moral crisis that famously conclude *Samson Agonistes*? Do these melancholic diagnoses amount to merely passing references, local details, a few drops of black bile that dissolve within the blood, sweat and tears of the larger work that surrounds them? What might a melancholy Samson contribute to the anxious problems of interpretation posed by Milton's text today? I hope to argue that this curiously extended series of references to melancholy casts a corrosively reductive physiological meaning onto the very "inward motions" upon which Samson's heroic final transcendence from abjection to martyrdom depends. If we are to risk parodic deflation, it may be that the flames that stoke the Phoenix-like "fiery Virtue" of Samson's self-sacrificial passions derive their source not from some powerfully exceptional moment of sovereign consciousness but stem instead from an overheated liver and a bloodstream clogged with the soot of burnt black bile.

Samson Agonistes has enjoyed a dubious vogue since the events of September 11th. Its already-formidable internal complexity and multivalence has only been magnified by the work's perhaps fatally overdetermined relationship of resemblance with what, until recently, one might still have termed our own contemporary theologico-political situation. As a mood of stasis that tends to refuse catastrophic resolution in favor of an endless malingering, melancholy seems to be the affect of choice that permits this temporal dynamic of "sticky" relevance between past and present to persist as we collectively refuse to stop mourning a dead object. Insofar as, according to the now-notorious Stanley Fish and John Carey contretemps, the interpretive status of Samson's inner motivation constitutes the Archimedean point upon which Milton's final literary testament can be said to either endorse or repudiate the religious politics of terror, the question of whether Samson would have been understood to be mentally ill by Milton's readers in the period provides some potentially crucial leverage in reorienting a text still obscured by its polemical collapse into the partisan and reductive terms of Bush-era theorizing.[2] Framed as a citation of a complexly valued and attacked stereotype, melancholy subjectivity always occasions a cascade of questions about the verification and evaluation of its performed feelings and symptoms: Are these emotions faked or real? Is markedly "humorous" behavior the expression of a personality or the symptom of a disease? Do passions necessarily produce failures in judgment?

Both belated as a vestigial survival of ancient medical theory and, in a "second death," already dated as a fashionable urban posture by the late seventeenth century date of *Samson Agonistes*' publication, melancholy's effects within Samson's mind and body would seem to pull Milton's work away from its current contemporary "relevance" to the era of suicide terrorism, instead chaining Samson to the deadweight of discredited quackery. But as Burton's textual practice demonstrates, melancholy is in some crucial senses an antihistorical phenomenon, a psychic stance of deliberate anachronism wedged against the forward progress of history, both personal and cultural, a mode that Elizabeth Freeman has termed "temporal drag."[3] Coined in relation to queer stances of obduracy or anachronistic slippage across the generational logics of cultural transmission, Freeman's phrase has gone viral in gay and lesbian studies, but it has also migrated toward early modern studies.[4] Inflecting the phrase melancholically, we might hear the word "drag" not simply as a temporal lag but in its emotionally idiomatic sense: drag as morass, malaise, bummer. Melancholic temporality functions here as the drag or travesty by which Renaissance subjects show up as Greeks to themselves, as a delay system by which the formerly fashionable becomes the humiliatingly passé, and as a quixotic means by which the self might subtract itself from the forward march of historical straight time. Widening the lens, one might also speculate that, by functioning simultaneously within the humoral physiology of classical Greek medicine and as melancholia within the psychoanalytic register of Freudian modernity, melancholy constitutes one of the privileged suture points joining together that hybrid monstrosity, the "early modern self" as an always already polychronic phenomenon.[5]

In pressing the question of Samson's melancholy, I seek to re-estrange the biblical Samson from both his Miltonic appropriation and his recent theologico-political reappropriation at the hands of Stanley Fish and John Carey. It also permits us to localize and complicate a broader issue within Milton studies: the ongoing dispute regarding his alleged metaphysical commitments to a substantial monism.[6] Entertaining the possibility of a melancholic Samson, in both the early modern humoral sense and in the contemporary senses of the word, might also help us to see an altogether different Samson than the strong silent type of Sunday school pageants: one who scans not only in the terms of early modern medical theory as both effeminate and melancholic, but also in the contemporary terms of gender theory and psychoanalysis as both masochistic and melancholiac.

If Samson was a melancholic, what sort of melancholic was he? The first step before attempting to characterize Milton's Samson as a melancholic would be to reassert, as I have been arguing throughout this book, that melancholy itself is not a unity but designates a broad taxonomy of social types within the period: Malcontents, lovers, courtiers, heretics, philosophers, and scholars all had their own specific forms of the condition, and there was correspondingly a wide spectrum of both valence (from the sobriety and genius celebrated by Florentine humanist Marsilio Ficino to the mania and psychosis treated by London physician Richard Napier) and salience (from passing moods to lifelong temperaments). This conceptual breadth keeps permanently open the critical question of whether Renaissance melancholy was a gift or a disease, a gentle sadness or a violent madness and, accordingly, whether melancholy can be understood as a material substance in the first place. Any given representation of emotional character within the period falls within a social signifying space framed by these generic typologies and scientific gearshifts, even as complex representations revise traditions and struggle to evade capture in their reductive terms.

But reduction has its pleasures, as Milton was well aware, and this marks the second step. Milton's definitive contribution to the literary representation of the humors, the companion poems "L'Allegro" and "Il Penseroso," opt decisively for a softened and aestheticized understanding of melancholy that furthers the "genial" melancholic tradition described in the introduction that stretched backward through Ficino to its roots in the pseudo-Aristotelian "Problemata XXX.1." The very narcotic smoothness of these poems, their coolly functional capacity to relay both titular moods in shared terms of pleasurable quasi-rural seclusion and labor-free contemplation, has been much remarked; and their success as mood-enhancing agents has led to these poems being more imitated and anthologized than analyzed. But these exercises (in every sense) from the 1630s should not be taken to exhaust and complete Milton's understanding of melancholy simply because they perform one specific distillation of humoral thought with bewitching finesse. Rather, Milton's very mastery of the genial melancholic repertoire of images and feelings suggests a rhetorical self-consciousness about the constructed nature of such heavily art-directed melancholy and, specifically, betrays in its very smoothness the substantial work of repression required in order to decisively purge genial melancholy of the physiological taint and intestinal odor of its explanatory rival: Galenic melancholy, which figured the condition not as a twilit reverie but as an unpleasant medical condition, a persistent state of cold, dry, and bilious stupor that induced depression and anhedonia.

Embodiment returns with a vengeance in the way that *Samson Agonistes* represents melancholy. The scholarly repose and vespertine music of "Il Penseroso" are conspicuously absent. Samson self-diagnoses his melancholic malady in a specifically physiological manner whose reference to vital spirits was no mere metaphor:

SAMSON: [. . .] my thoughts portend
That these dark orbs no more shall treat with light,

> Nor the other light of life continue long,
> But yield to double darkness nigh at hand:
> So much I feel my genial spirits droop,
> My hopes all flat, nature within me seems
> In all her functions weary of herself;
> My race of glory run, and race of shame,
> And I shall shortly be with them that rest. (590–598)
>
> MANOA: Believe not these suggestions, which proceed
> From anguish of the mind and humours black,
> That mingle with thy fancy. (599–601)

To be perfectly Galenic, Manoa means that thick black juices from overheated blood are producing vapors that cloud Samson's mind. It can be difficult to take this kind of medical plumbing talk seriously, in our rush to regard "anguish of mind" as an intuitively plausible notion, because it correlates with the historical horizon of our own psychological comfort zone while regarding "humours black" as a vestigial trace of outmoded science. Such unbecoming anachronisms jangle harshly with the dignity of Samson's indignity, so to speak. It is as if Clytemnestra were to burp mid-sentence and complain of tragic heartburn. But we would do well to recall that multiple physiological and philosophical understandings of emotion would have been available to Milton, comprising both evidence for and against humoral medicine, and evidence for and against competing understandings of melancholy. In the moment of engagement, the emergent winners and recessive losers of intellectual history were not clearly marked as such. A comparative case in point: Rene Descartes' 1649 text *The Passions of the Soul*, in which Descartes approvingly cites Harvey's theory of the circulation of the blood and yet notoriously locates the soul's material interface with the body in the pineal gland.[7] We would do well not to expect early modern thought on the mind and the body to anticipate our own in any straightforward manner.

That Milton did not necessarily reject Galenic understandings of the physical support systems that could reinforce, if not produce, affect, is implicit in the dialogic sequence, and it is not simply specific to Manoa. No sooner has Manoa's detection of black humors been presented, then Samson himself launches into his longest set-piece of melancholic self-representation, one that is worth quoting at length:

SAMSON: O that torment should not be confined
To the body's wounds and sores
With maladies innumerable
In heart, head, breast, and reins;
But must secret passage find
To the inmost mind,
There exercise all his fierce accidents,
And on her purest spirits prey,
As on entrails, joints and limbs,
With answerable pains, but more intense,
Though void of corporal sense. (606–616)

The word "spirits" here does not designate the ghostly or the immaterial but the finest and most subtle form that material embodiment can take: The animal spirits are the most refined liquid messengers within early modern living systems, emissaries between the brain and heart and musculature that trigger memories and produce motor responses. Registering the narrowest infra-thin boundary at which a living system can suffer from the alienated disconnection of Cartesian dualism, in flagging that even these minuscule physical spirits can become the victims of emotional torment though they are "void of corporal sense," Samson here laments the intuitively obvious fact that mental pain and bodily distress interrelate and mutually interpenetrate. What interests me here is not only that inwardness and femaleness are both imagined together in opposition to a masculine torment which exercises "his" accidents upon "her" entrails. Rather, this sexed morphological imaginary of active male and passive female figures a more radical and more mysterious interaction between opposed terms: the bodily and the incorporeal.

Samson's self-identification as melancholic hinges upon metaphorical equivalences between the psychic and the somatic whose specific tenor/vehicle relationship exploits the flexibility of early modern understandings of what Lawrence Babb foundationally defined as a "physiological psychology."[8] Such language is thus not *merely* the metaphoric joining of separate categories but a reference to a historical horizon of knowledge in which the two were understood to be versions of each other linked in resonant microcosmic/macrocosmic correspondence. In the *Ethics*, surely the most specialized of all cases when considering early modern models of mind/body interaction, Spinoza defines melancholy as an affection which is equally present in the body and the mind, in contradistinction to the primarily mental state of Anguish.[9] Nor is Samson alone within Milton's text in connecting thinking and feeling in this manner. His remarks here remind us that the chorus, too, read psychological pain as physical injury, and imagine earlier in the text that their comforting words are a kind of medical intervention: "Salve to thy sores, apt words have power to suage, / The tumors of a troubled mind, / and are as balm to festered wounds" (184–186). What is ironic in their case (for the chorus is hardly reliable or comforting in their remarks, and sometimes verge upon playing "miserable comforters" to Samson's Job) nonetheless illuminates a conceptual backdrop shared by all the characters within the text: the sense that the mental and the physical are to be understood as intimately bound together in mutually causal reciprocity.

This general backdrop becomes character-driven in the way that Samson implicates himself through his accounting of not only how but, at least indirectly, *why* his "genial spirits" are drooping. Physiology replicates autobiography. In bewailing the possibility that even the "purest" spirits can be contaminated by the secret inward passage of melancholy, Samson narrates his own sorry personal history as a Nazirite hero who has allowed a heathen woman to have access to his inmost secrets and thus betrayed his people. That is, Samson's melancholy and his masculinity must be thought together in multiple dimensions as modalities that figure each other: In a narrative logic of exchange, the sexual shaming implicit in Samson's capture, blinding, and shaving is both the occasion for the destruction of his masculinity and a priming condition or vector for his melancholy. Crossing over from active warrior to passive prisoner, from hairy hero to shaved slave, the collapse of normative masculinity produces melancholy. It is not that masculinity is a thingly unity which one can simply have and then lose. Rather, as one's gendered self-understanding undergoes change over time, the self-satisfaction accrued by successful and unsuccessful performances of masculinity produce attendant praise and blame whose lived affect can thicken or dissolve the internal consistency of melancholic substance within the body. This interior oscillation with respect to tenuously grasped gender positions that both abide within a single subject complicates the dispersal of "light" and "heavy" melancholy between men and women previously seen in "Love's Labour's Lost"; however much he has been associated—rightly or wrongly—with patriarchal agendas, here Milton thinks gender from the inside as a laborious, fraught, and incomplete process of attenuation and change.

"O blot to honor and religion": Masculinity and Effeminacy

As with the suggestion that Samson might be understood as a melancholic, the allegation that there is something compromised about Samson's masculinity might initially seem simply perverse. Who could be more masculine than Samson? Samson's legendary feats of strength before his capture and his stony, stoic refusals to engage with his father Manoa, his lover Dalilah, and his rival Harapha, and, finally, his ultimate act of superhuman athleticism in the destruction of the feast of Dagon, together consolidate his identity as the definitive avatar of masculine fortitude. Far from impregnable and self-evident, in *Samson Agonistes* Samson's manhood is marked first and foremost by a persistent anxiety about its capacity to betray itself and transform into a disastrously compromising effeminacy, one whose verbal description and denunciation constitutes a self-flagellating refrain that is integral to the maintenance of Samson's melancholy feeling. In order to argue that *manhood* and *masculinity* are the neighboring gendered terms against which Samson's *effeminacy* shows up as a threatening structural inversion, I must first flag a historical distinction about which of these norms can be said to be in play. As Amanda Bailey has noted,

> The word 'masculinity', which did not enter the English language until the middle of the eighteenth century, referred to the privilege awarded to men in matters of inheritance. 'Manhood' and 'manliness' were the terms used in the sixteenth century to connote those qualities essential to civility, which was identified teleologically as the definitive characteristic of the adult man.[10]

In this sense, "manhood" should be understood not only in relation to boundaries of gender but also to boundaries of ethos, polis, and species; in early modernity, the opposite of manhood is not only "womanhood" or "femininity" but also in-civility, brutality, animality, in-humanity. The conceptual space of the "un-manly" thus constitutes a negative reserve in which class, ethnicity, species, and gender differences mutually figure one another, and reservoirs of meaning from any of these separate registers can flow into the space opened up within manhood by effeminacy.

Anxiety about the inward contamination of male gender by effeminate possibility constitutes a thread of queer fear that runs throughout Milton's poetry and his prose, taking different local forms each time, but sharing a family resemblance with its locus classicus, the angelic reproach in *Paradise Lost* directed at "Man's effeminate slackness" (11.632). Samson's self-hatred focuses its energy through an insistent proclamation of the "effeminacy" he supposedly demonstrated in succumbing to Delilah's demands that he reveal the secret of his strength:

> At times when men seek most repose and rest,
> I yielded, and unlocked her all my heart,
> Who with a grain of manhood well resolved
> Might easily have shook off all her snares:
> But foul effeminacy held me yoked
> Her bond-slave. O indignity, O blot
> To honor and religion! Servile mind
> Rewarded well with servile punishment! (406–413)

Samson's scenario of "yielding" and "unlocking" in the night summons up the remorseful laments of despoiled maidens bewailing the loss of their virginity—but ironically what has been lost here (momentarily displaced? forever dissolved?) is not maidenhood, but "manhood." The reference to other men as a class marks Samson as somehow cut off from homosocial solidarity, defenseless against his own servile gullibility. "Holding" him against his will, effeminacy is made here into an agent whose foul embrace cannot be withstood.[11]

Here we must attend to a discrepancy in the signification of *effeminacy* itself within the period. The term could designate a male with "womanly" characteristics, its first meaning, but it could also signify a male with an inordinate weakness *for* women. As the *OED* notes in reference to usage in Caxton (1460) and Puttenham (1589), "the notion 'self-indulgent, voluptuous' seems sometimes to have received a special coloring from a pseudo-etymological rendering of the word as '*devoted to women*'. Unequivocal instances are rare."[12] If we keep this second definition in mind, then Samson's self-accusation may simply be directed at his gullibility, and his indulgence of his wife. Certainly the Biblical source in the book of Judges sounds this note, with its joke-like repetition of Delilah's demand for Samson's secret eventually producing the desired revelation. Secondarily speaking, "effeminate" could simply be a synonym for "uxorious."

Yet the capacity of the meaning of effeminacy in the period to slide between its two distinct definitions also usefully marks an ambient difference between early modern understandings of gender and normative (if equally debatable) "modern" schemas of sexual orientation. In our own cultural moment, in which we are subject to the reified sexological categories of heterosexuality and homosexuality and are schooled by popular psychology to probe for the occulted undersides of how those locations manifest themselves in everyday life, an "effeminate" male potentially falls under suspicion of so-called "latent" or unacknowledged homosexuality in a manner compatible with the first meaning but necessarily incongruous with the second. By contrast, for early modern subjects the polarities of "masculine" and "feminine" stand in a more volatile relationship as, on the one hand, social positions structured by rigid and divinely ordered prescriptions about rule and obedience, and, on the other, developmental outcomes placed by classical physiology into a rather slippery and tangled proximity.

Effeminacy's second definition within early modernity draws its strength from a historically distinct morphological imaginary sourced, ultimately, in a classical inheritance. Samson's sense of effeminacy as an invasion from within recalls John J. Winkler's formulation of sex and gender in classical Mediterranean culture as one in which " 'woman' is not only the opposite of man; she is also a potentially threatening 'internal émigré' of masculine identity."[13] Such a possibility might be said to constitute the bad dream of the so-called "One Sex model" so widespread within New Historicist readings of Galenic physiology in the wake of Thomas Laqueur's seminal *Making Sex*. If women are only born women because of their stalled developmental progress within the womb en route to becoming men, might it not also be possible that "from within" a man might somehow lapse and slide backward toward another gendered location? This is the inverse scenario to the hermaphroditic possibility that animates that touchstone of New Historicist thought on the swerves and curves of gender normativity, Steven Greenblatt's "Fiction and Friction" in *Shakespearean Negotiations*, in which sufficient heat permits the extroversion of "Marie" le Marcis's female genitals into their final, normative form as the male genitals of Marin le Marcis.[14] If the pseudo-revolt of transvestite theater described therein through Greenblatt's reading of Viola occurred under the protective shade of masculine superiority, then, framed against this New Historicist critical rubric, the effeminate male is the early modern gender system's worst possible outcome. This is not because he is a figure neither successfully masculine nor authentically feminine (granting that such secondary authenticity is in a peculiar sense impossible given the inherent insufficiency of the feminine position) but because, as a backslider, he is the only agent capable of betraying the forward course of masculine supremacy itself. Loitering with intent in a contaminating interstitial space between genders, the early modern effeminate man is a gender recusant.

Acutely afraid of his own effeminacy but also eager to self-consciously punish himself for it, Samson seems at pains to accuse himself of specifically the second kind of effeminacy—but he risks protesting too much in the process, and his accounts of that condition slide uncomfortably "forward" (proleptically, historically) toward the first definition, with its connotations of a contaminating inward marker of feminine qualities and, specifically, with a feminine "weakness" in the face of sexual advances. In Samson's imagination, this state capaciously opens itself to suggestions of anal rape and military dramas of subjection, captivity, and male homoeroticism. The self-accusation of "effeminacy" recurs in Samson's thoughts on the uselessness of a merely occasional temperance in a manner which telegraphs his terror at occupying a permissive, receptive, passive position: "What boots it at one gate to make defense, / And at another to let in the foe / Effeminately vanquished?" (560–562). The "other gate" in this image is an unexpected entryway into one's self. In this military scenario (the image is one of attackers penetrating a town's line of defense), "to be conquered" and "to be effeminate" somehow lead to and verify each other. Effeminacy makes one a ripe and justified target for conquest, and having been conquered confirms and ratifies that effeminacy, securing a lasting shame for the vanquished by retroactively projecting a prior condition felt to somehow merit domination as its confirming response.

Voiced in provocative and dramatic terms seemingly intended to exacerbate his present dejected mood, Samson's self-castigating acts of autodescription as effeminate have the effect of recharging his psychic battery with fresh affective currents of pain and humiliation. I shall argue in a moment that this reflexive habit constitutes a kind of belligerent truthfulness of self-description that aligns Samson's early modern melancholy with the modern psychoanalytic category of melancholia described in *Mourning and Melancholia*, but first I wish to point out that, insofar as these self-denunciations produce some modicum of self-satisfaction in performing suffering and abasement for others, they also exhibit traits that belong to another psychoanalytic category: masochism. Bracketing the theoretical disjunction between melancholia (theorized at the level of libido) and masochism (theorized at the level of the drive) for psychoanalytic knowledge today, what connects Samson's masochism with his melancholy within Milton's text?

"Choosing Death as Due": Samson as Masochist

The possibility that Samson somehow desires suffering rather than merely endures suffering first arises in his instantaneous rejection of his father Manoa's attempts to negotiate a ransom for Samson's release ("Spare that proposal father, spare the trouble / Of that solicitation, let me here, / As I deserve, pay on my punishment" 487–489), a response whose slightly overstated repetition registers the stirrings of deep feeling as an insistent redoubling echo within the syntax.

Manoa prods at the feelings that underwrite this response by drawing a distinction between bearing one's allotted due of suffering and willfully seeking to extend it needlessly: "Be penitent and for thy fault contrite, / But act not in they own affliction, son" (502–503). The paternal injunction to "act not" registers at a micrological level a surrounding climate of rhetorical associations between the passion of martyrdom and the affectation of theatrical acting. In her essay "Aesthetics as critique: Tragedy and Trauerspiel in *Samson Agonistes*," Victoria Kahn has explored this set of links between acting and martyrdom as a problem for revolutionary political projects: It is part of a revolutionary project to repudiate the sympathy that affected displays of suffering seek to produce.[15] The lure and threat of such sympathies clearly troubled Milton, who did not hesitate to attack the histrionics of Charles I's martyrology in a manner that constructed both the decapitated king and his posthumous cult of mourners as effeminate and affected.[16] The paternal voice, in reminding his son of the difference between sincere penitence and "acting," schools him in masculine fortitude precisely at the moment in which Samson would seem to be demonstrating it most publicly. But that very publicity is the problem with "acting" one's affliction rather than simply enduring it.

Manoa repudiates the directness of an explicitly masochistic decision to suffer in favor of trust in the possibility of divine mercy:

> Who evermore approves and more accepts
> (Best pleased with humble and filial submission)
> Him who imploring mercy sues for life,
> Than who self-rigorous chooses death as due;
> Which argues over-just, and self-displeased
> For self-offence, more than for God offended. (510–515)

Manoa detects, in short, a kind of wounded narcissism in Samson's gluttony for punishment, a sense that only an exaggerated self-regard would insist upon gratuitous displays of suffering in response to the shame of Samson's compromise and capture. Rather than an expression of humility, the decision to effectively seek out a surplus of suffering constitutes a kind of negative afterimage of pride.

In a response that seems to exemplify, rather than disavow, Manoa's charges of "self-regard," Samson recounts to himself his glory, wielding his exemplary strength and status as a tool with which to further bewail his currently debased and abject condition:

> SAMSON: Fearless of danger, like a petty god
> I walked about admired of all and dreaded
> On hostile ground, none daring my affront.
> Then swoll'n with pride into the snare I fell
> [. . . .]
> At length to lay my head and hallowed pledge
> Of all my strength in the lascivious lap
> Of a deceitful concubine who shore me
> Like a tame wether, all my precious fleece,
> Then turned me out ridiculous, despoiled,
> Shaven and disarmed among my enemies. (529–532; 535–540)

The snare into which he falls is, allegedly, one fashioned by the "fallacious looks" and "venereal trains" of foreign women. But pride seems to continue to register in these backward glances at his own lost image. Though it cannot be established with any certainty, it is strongly tempting to hear in the mingled notes of melancholy and masochism in this exchange between Samson and Manoa some strong echoes of another melancholic whom we have already examined, Shakespeare's Antonio from *The Merchant of Venice*. This comparison might seem merely arbitrary and fanciful were it not for a peculiarly apt combination of drives and images: Faced with the imminent threat of Shylock's harvesting of the pound of flesh in the trial scene, Antonio seems to rush headlong past the ongoing rescue efforts of Bassanio and others, and all-too-eagerly cries "let me have the judgment" (4.1.83), and, in the very same scene, Antonio describes himself as a "tainted wether of the flock, / meetest for death" (4.1.114). Both Antonio and Samson figure themselves to others as castrated rams, a histrionic act of abusive self-description that reproduces the shame it describes. Both Antonio and Samson seem in a larger sense to desire the spectacle of their own punishment, to produce it as a legible scene for a wide audience. With a distinct eagerness, Samson imagines "oft-invocated death" as "the welcome end of all my pains" (75–76). Samson's self-presentation as a melancholic frozen in a quasi-catatonic shutdown of despair yields then to a masochistic embrace of suffering as a form of heroic passivity which both transcends and preserves melancholy affect. That is, Samson's capture signifies for him a gender failure which occasions both melancholy and masochism, but also multiplies each by the other: The sadder and more passive Samson becomes, the more he "suffers" his melancholy—but the more he suffers, the more he transforms melancholic suffering into a kind of masochistic public enjoyment of his dejected/enslaved/shaved/blinded status. The humiliating legibility of that status to others earns him a kind of moral credit within a divine calculus of suffering. The more he denounces himself as effeminate and worthless, the lower he becomes in the estimation of his social surround. Yet by lowering himself before anyone who will speak to him, he demonstrates and ratifies his repentance in a manner designed to elevate him in the eyes of his intended audience, God.

Accustomed to a psychoanalytic treasure hunt for the secret pressure of castration anxiety within everyday performances of masculinity, we are perhaps left oddly ill-prepared for its eruption in broad daylight as manifest content. When a male character openly describes himself as castrated, we may need to slow down and seek to place this image in the concrete specifics of its religious and cultural connotations. Samson's image of himself as a wether collapses his consecrated status as a Nazirite dedicated to God, his betrayal of that sacred mandate, and the subsequent punishment (shaving and blinding) together into a densely overdetermined sexual and religious shorthand. "Overdetermination" may sound rather pat, but it suits the pileup of registers here. Does a religious symbol of the sacrificial animal code a sexual subtext? Or does a sexual image of castration code the religious shame at allowing sacred masculinity to be despoiled and destroyed by the pagan enemy? Obviously, seeing the sexual and the sacred *in flagranti* was no scandal in early modern religious discourse: From the "Song of Songs" to Donne and Herbert, this modular interdependence proves noticeably generative in the religious poetics of the period. The overdetermination in play within the "wether" image is not simply the redoubling of Samson's sexual shame and his religious shame, but the mutually reinforcing position it occupies between Samson's masochism and his melancholy as interpretive problems.

In the conclusion to her *Habits of Thought in the English Renaissance: Religion, Politics, and the Dominant Culture*, Deborah Shuger challenges the usefulness of masochism as a paradigm for the interpretation of early modern psyches. She does so with reference to Donne, but in a manner that raises important questions that also challenge my own use of masochism here to discuss the literary creations of Milton and Shakespeare:

> The public character of language guarantees that the relation between "inside" and "outside" more resembles a Mobius strip than nesting boxes. This is why one cannot translate something like Donne's valorization of guilt as masochism; the experience qua experience has already been interpreted, so that to alter its name is to alter its essence. We might postulate a primal feeling—the need to be punished—but the nature of the experience depends upon whether that feeling is interpreted as redemptive or pathological.[17]

First, I want to suggest that it is exactly psychoanalysis that is in the best position to deliver upon the topological metaphoric relation between language and the self that Shuger here invokes via the Mobius strip. The notion of a hermetic and absolute divide between an "inside" and an "outside" is not how psychoanalysis conceives of the source of structure within the unconscious, though Shuger's critique seems to inadvertently describe her own opposition between "theory" (abstract, general, static) and "culture" (material, specific, historical) rather well, and certainly better than it describes psychoanalytic thought. More importantly, I wish to reject Shuger's "either/or" articulation of redemption and pathology, but instead would counter that Samson's perverse and ambivalent "rousing motions" are themselves structured by Milton to show up for us as "both/and": They force our experience of a simultaneity which is both redemptive for Israel in its consequences and, in its expression of melancholic despair and violent self-hatred, resolutely pathological *at the same time.*

Pathology and/or Redemption

The double exposure of the redemptive and the pathological is, in fact, another way of expressing what I have taken all along to be the constitutive outcome of the early modern archive of melancholy. Early modern culture understood melancholy as the site of an endless simultaneity between the redemptive (Aristotelian genius, Platonic "frenzy") and the pathological (Galenic illness). And this "both/and" characterizes the Mobius strip, or auto-differential loop in place between melancholic self-hate and masochistic self-harm. The two are distinct, and yet they feed forward into each other in a self-recursive assemblage. Samson sees himself as effeminate, and in hearing how he uses this self-accusation we can also hear him as a masochist. These effeminate and masochistic frameworks for self-presentation function as paradoxically pleasurable postures through which Samson expresses his melancholy. Melancholy is the "other" side of the masochistic Mobius strip. But in order to see how that works, we need to understand Samson as not simply melancholic in the early modern sense but as a melancholiac in the Freudian sense.

Samson's self pitying complaint commences Milton's text, and if its oft-quoted *cri de coeur* risks overfamiliarity, it is nonetheless essential to establishing early the circularity of his despair: a one track mind, both monotonous and rhythmically repetitious: "O dark, dark, dark, amid the blaze of noon, / Irrecoverably dark, total eclipse / Without all hope of day!" (80–82). Note that this evocation of "darkness at noon" faintly recalls the noonday demon of despair spoken of in the Psalms, and described by Giorgio Agamben in *Stanzas* as the root of monastic acedia, the sinful despair of the possibility of salvation, an inner shadow of self-judgment which Samson here telegraphs as "hopeless" and "irrevocable." Cut off by blindness from visual access to the light that would signify God's "prime decree," Samson remains subjected to an unseen human world of misery and defeat. The result is a kind of writhing proprioceptive shame at his own embodiment, whose sightlessness degrades him below the lowest insect life forms, and whose conditions of survival are constantly mediated by concession to unseen but omnipresent enemies:

> As in the land of darkness yet in light
> To live a life half-dead, a living death,
> And buried; but O yet more miserable!
> Myself, my sepulcher, a moving grave,
> Buried, yet not exempt
> By privilege of death and burial
> From worst of other evils, pains and wrongs,
> But made hereby obnoxious more
> To all the miseries of life,
> Life in captivity
> Among inhuman foes. (99–109)

Samson's sense that his "self" is his "sepulcher" turns the tautological circuit of self-identical ipseity (I = I) into a feedback loop of shame at the conditions of his very survival.[18] Relentlessly confirming and reinforcing itself, this cycle recalls Proposition 55 from Book Three of Spinoza's *Ethics*: "When the mind thinks of its own impotence, by that very fact it feels pain."[19] To continue to live is to continue to accept the terms of his captivity and domination, and so the "impotence of mind in body strong" that Samson bewails indexes both the subjection of a strong body to an impotent mind and the impotence of the mind itself to think its way past despair. As long as he is "still here" within his blinded and subjected body, he is still trapped in a life/death that offers neither the autonomy of genuine life nor the security from harm of genuine death. Survival connotes submission.

So why identify such reflexive woe-is-me summarizing as melancholia rather than other handy contemporary affects such as shame or guilt? It is not that there is something deluded or false about Samson's précis of his situation; in fact, his words may be all too accurate. But it is the pronounced relish with which he rehearses his flaws and faults, the remorselessness of his very remorse, if you will, that flags his relationship to despair as linguistically productive in the kind of telling manner that Freud, with reference to Hamlet, described in "Mourning and Melancholia" as a specifically melancholic glee in self-denigration: "The patient represents his ego to us as worthless, incapable of any achievement and morally despicable; he reproaches himself, vilifies himself, and expects to be cast out and punished. He abases himself before everyone and commiserates with his own relatives for being connected with anyone so unworthy."[20] Note especially that in Freud's account, truthfulness itself takes on a kind of melancholic glamour, an ostentatiously strict accounting of the self's wrongdoing whose unvarnished purity strobes between the positivity of an affirmation of what one has done and the negativity of a stripping away of any of the standard cushions of comforting self-esteem:

> When in his heightened self-criticism he describes himself as petty, egoistic, dishonest, lacking in independence, one whose sole aim has been to hide the weaknesses of his own nature, it may be, so far as we know, that he has come pretty near to understanding himself; we only wonder why a man has to be ill before he can be accessible to a truth of this kind. For there can be no doubt that if anyone holds and expresses to others an opinion of himself such as this [. . .] he is ill, whether he is speaking the truth or being more or less unfair to himself.[21]

The brutality of Samson's honest appraisal of himself as insect-like, bound for death yet unable to die, exposed daily to endless humiliation: It is the vicious surplus of honesty characteristic of this sort of self-scrutiny that, for Freud, tunes melancholic selfhood at a deliberately sharp pitch. To rehearse Freud's titular dyad, melancholy's definitive difference from mourning consists in the refusal to permit decathexis. The withdrawal of emotional investment from a lost object typical of normative mourning is replaced in melancholy by an obstinate eagerness to keep the wound open and on display for others through a bewilderingly aggressive show of self-hatred. Normative mourning allows for a reality testing in which emotional investment in a lost object is gradually subtracted; through a kind of structural mimicry, Freudian melancholia by contrast involves not the genuine loss of somatic death but the emotional loss of an object of love, an object which is encrypted into the self and which obstinately persists there as an undead remainder. Equating "my self" and "my sepulcher," Samson announces the melancholiac mourning of his own self as a lost object of love.

"Rousing Motions": Aristotle, Galen, and the Materiality of Emotion

How can we peer with any security into the irrecoverable darkness of this inner grave? Does Samson's present melancholic self mourn his lost masculine self or, pressing further within, his lost effeminate self? These problems of access and verification are the primary effect of melancholic representation within the period, insofar as access to humoral insides was always routed through guesswork at symptomatic outsides, guesswork with no guarantees, hedged by the risk of modish insincerity on the one hand and mere madness on the other. Is melancholy the demonic and parodic opposite of sincere religious penitence, or is melancholy precisely the affect of sorrowful self-reckoning and inward vision that indexes sober reflection and moral seriousness? This hermeneutic of mystery only intensifies as Milton's text approaches, and works through, its awful climax: At the conclusion of the text, Samson finally decides to allow himself to be exhibited by the Philistines at the feast of Dagon and, in a sudden and mysterious act of sacred violence in which his divine strength resurfaces, the shamed prisoner pulls down the pillars of the theater upon himself and the assembled foreign court. Tracing the arc of Samson's heroic transcendence from abjection to martyrdom depends upon how we evaluate his final action, but his final action is said to derive its meaning from the motivations that produce it, and those motivations are described in a manner that pulls us toward the edge of a certain epistemological abyss whose contours are the diagnostic contours of melancholy: the question of whether inner psychic pain is a mark of divine inspiration or a symptom of private pathology. The standard opposition in *Samson Agonistes*' criticism, and the question at the heart of the Stanley Fish and John Carey debate (do the "rousing motions" Samson feels come from God or from Samson's own deluded psyche?) repress a materialist third option: The motions come not from the melancholic mind but from the melancholic body, a source for actions and consequences that would have been entirely plausible to an early modern subject but which we feel less comfortable with today. What if the rousing motions are the circulation of humoral fluids and the flow of animal spirits across the subtle membranes joining Samson's mind and body? The same morphological imaginary which figured the process of "black humours" doing their destructive work upon the "secret passages" of Samson's "inmost mind" persists, albeit in a positive valence, in the language of inward movement that drives the "rousing motions" speech:

SAMSON: Be of good courage, I begin to feel
Some rouzing motions in me which dispose
To something extraordinary my thoughts. (1381–1384)

How are we to understand this simple phrase, "rouzing motions"? What philosophical or physiological understandings stand behind this bare pairing? Motion itself produces intolerable puzzles for those committed to thinking being as essence, and resists even thinkers with relatively hearty accounts of how beings can undergo change. The Aristotelian philosophical tradition Milton would have absorbed at Cambridge offers at least a plausible starting point, and we can see in the *Metaphysics* that Aristotle acknowledges the difficulty in conceptualizing kinesis:

> Movement is supposed to be a sort of imperfect actuality, for the reason that the potentiality, whose actuality it is, is incomplete. And therefore it is hard to grasp what movement is; for it must be classified either under "privation" or under "power" or under "pure actuality" and none of these appears to be possible. Consequently, what has been said remains plausible: movement is a kind of actuality which has been described, which is hard to discern, but which is capable of being.[22]

It is striking that in working through even this admittedly guarded and tentative account of kinesis, Aristotle invokes not only differences between health and sickness but specifically changes in the state of the blood: "To be capable of being healthy and to be capable of being diseased are not the same (if they were, being healthy and being diseased would be the same). But it is the same subject that may be either healthy or diseased, such as blood or some other bodily fluid."[23] Motion is imagined here in relation to changes of inward states, and at least minimally suggests an Aristotelian precedent for thinking about inward movement or "rouzing motions" in relation to the evaluation of transformations from sickness to health and back again, rather than in terms of emotional and passionate mental stances for a consciousness. What prevails in Aristotle is a material and substantial account of motion as change—but does this necessarily determine Milton's own thinking? As N. K. Sugimura has pointed out in the conclusion to her account of "matter" in Milton's texts, "The terms 'substance' and 'matter' elicit in his poetry a metaphysic that breaks free from the older, predominantly Aristotelian conceptions to which he was tied hitherto, but which it nevertheless required in order for thought to proceed."[24] The question remains: Breaking free toward what?

Bracketing this possible but by no means assured Aristotelian background, given the tenor of Samson's masculine credo of endurance and sufferance, I expect that Stoic philosophy might also underlie this account of the status of inward "motions," and Stoicism constitutes a likely ethical backdrop that supplements Hebraic heroism with a no less masculinist classicism. Galen's *On the Doctrines of Hippocrates and Plato* contains a kind of miniature account of Stoic ideas about the passions which seem to convey the sense in which "motions" could designate a pejorative and irrational disturbance of the mind away from reason, though we must be cautious given the secondary nature of Galen's summarizing:

> In book I of his *On Passions* Chrysippus tries to show that the passions are certain judgments [formed] by the reasoning part of the soul, but Zeno did not believe that the passions were the judgments themselves but contractions, relaxations, elations, and depressions which supervene on the judgments. [. . . .] [Chryssipus] is in conflict with Zeno on this point, and with himself, and with many other Stoics who did not suppose that the passions of the soul were judgments of the soul but rather the irrational contractions, depressions, 'bites' elations and relaxations which follow those.[25]

It is perhaps in this sense that the idea of a "motion" within the soul occupies a place of curious abeyance on the cusp of definition in terms of its valence in directions that could be alternately demonic and divine, sacred or sick. For Zeno and Chrysippus (as filtered through Galen), "motions" is inherently elastic in its status, and so we do not encounter a resolving ground for the aporetic issue at hand in reading Samson's inner experience as either an elation or a depression.[26] But the Stoic account is nonetheless helpful in indexing motion *as such* as a straying from the firm baseline of an allegedly immobile reason.

Whatever our sense of its philosophical background, the result of the rousing motions is a plan of action, which Samson then glosses in a manner that strengthens the speculative hold of melancholic meaning:

SAMSON: I with this messenger will go along,
Nothing to do, be sure, that may dishonour
Our law, or stain my vow of Nazirite.
If there be aught of presage in my mind,
This day will be remarkable in my life
By some great act, or of my days the last. (1384–1389)

However bitterly divided Milton's readers remain about the "greatness" of the massacre itself, the day in question *is* both Samson's last day and the day of his "great act." Does Samson see his own future, and would that make him more or less melancholic? The association of melancholy with prophetic power within the period is discussed in detail by Winfried Schleiner in his classic study *Melancholy, Genius and Utopia in the Renaissance*, but for our purposes it is simply worth repeating that such associations are themselves sharply divided between the positive and negative polarities of melancholy itself: For those of the "genial" melancholy school, successful acts of precognition would further establish the divine favor of inspired melancholy that connects Platonic "frenzy" and Aristotelian genius together as modes of visionary perception. For those who regard melancholy as a form of pathology, false claims to prophetic powers would be of a piece with the generally skeptical attitude toward "enthusiasm" in the period. In particular, anti-Puritan writings often flagged the dangerous delusions of mentally ill rabble-rousers whose appeals to inward inspiration were notoriously impossible to verify or gainsay.[27] Milton courts the possibility that Samson is inspired with some intuition of future outcomes, but his final farewell that whether this day is "the last of me or no *I cannot warrant*" (1426; italics mine) bars any positive affirmation of precognition beneath the cover of humility.

Arguing that the rousing motions refer to the inner liquid flows of melancholy does not, in and of itself, resolve the question of how we are to evaluate and judge Samson's actions, for melancholy itself remained the contentious site of rival traditions and counter-traditions. Milton could be either endorsing or critiquing Samson by using various aspects of the melancholic tradition to serve either purpose: As the equivocal cultural status of melancholy demonstrates, there were melancholy heroes to be found in Aristotle and melancholy fools to be found in Galen. As ever, Galenic and Aristotelian variants of melancholy determine the broad options of interpretation, but the quirk of personal character evades their generic closure—yet this kind of interpretive frustration, this nagging refusal to abide on either side of the healthy/sick division, might itself be the keynote of a second order form of melancholy, an epistemologically "soft" melancholy that reveals itself as the very refusal to scan, the melancholy of an affective illegibility.

There is a more than accidental relationship between the affect of melancholy and the epistemological structure of aporia within early modernity. The representation of melancholy as a liquid that may (or may not) circulate within a substantial body and produce outward signifying effects stands in a mimetic relationship with the spatialized epistemology of artistic, dramatic, and formal scenarios that veil, encrypt, or conceal secret knowledge "within" a mysterious outer structure. From basic childcare to the philosophy of mind, incomplete access to the emotions of others constitutes one of the essential conditions through which problems of knowledge are worked upon, from both directions; the question of how to interpret the self from outward signs becomes particularly acute in the early modern period when the question is how to determine whether someone, one's self included, is or is not melancholic. Working from the outside in, Gertrude and Claudius seek to discover whether Hamlet is mad, and if so, why. Working from the inside out in the wake of the cogito, Descartes' criteria of clear and distinct conception is itself a negative purging of the dark and murky phantasms of melancholic visionaries, figures that stand in close textual and philosophical proximity to the posture of radical doubt. Six years later, he has barely begun the *First Meditation* when he pauses to consider whether he is simply suffering from melancholic delusion.[28] Melancholy and aporetic limit-cases of interpretation go together, and Milton's Samson belongs among these cases, for Samson's self-as-sepulcher figures the self not as a shamefully public place of exposure open to all but as an absolutely private zone, a radical and complete unknown for the "others" who blindly surround him. By the end of the poem, that self-as-sepulcher is crucially situated within a mass grave, a melancholy assemblage of corpses memorably evoked in Gordon Teskey's chapter "Samson and the Heap of the Dead" from *Delirious Milton*. Teskey argues that "the impact of the heap, which touches on the extremes of horror and exultation, begins in the spectacle of the other as phenomenologically uncertain, which in the realm of texts means *having no literal sense*."[29] But it is the "rouzing motions" within Samson himself which constitute the center of phenomenological uncertainty here, for they remain steadfastly, stubbornly *hors du texte* even as they are indicated within it. Samson's "inward" motions are just as radically offstage as the genuinely offstage catastrophe at the feast of Dagon, just as far beyond the reach of scrutiny, and in placing them there Milton relocates Samson from the exegetical framework of a Biblical heroism predicated upon vengeance and into a specifically early modern representational paradigm of melancholic aporia and aporetic melancholy.

All Passion Spent?

With the noted exception of Teskey, recent Milton studies have been allergic to aporia, and not simply in a reflexive reaction to its surplus ubiquity during the heyday of deconstruction.[30] Declaring with characteristic absolutism at the close of *How Milton Works* that "everything that many readers find interesting in Milton's work—crises, conflicts, competing values, once-and-for-all dramatic moments—proceeds from error and is finally unreal," Stanley Fish has alleged that any apparent conflicts, debates, or antagonisms that we perceive within Milton are merely indices of our failure to grasp the unity of Providential design from the standpoint of a post-Enlightenment liberalism.[31] Turning down the temperature slightly, one can sense the same resistance to aporia at the conclusion of John Knott's chapter on Milton and martyrdom from his comprehensive survey *Discourses of Martyrdom in English Literature, 1563–1694*:[32]

> Unless we are to view Samson as an ambiguous or perverse figure at the end of the play, both unlikely explanations in the context of Milton's work as a whole and his late poetry in particular, I believe that we must regard his destructive act as a display of divine vengeance and divine power, made possible by Samson's willingness to act as an instrument of God.[33]

This tense, counterfactual "unless" wards off the threatening specter of perversity, securing a reassuringly unambiguous Milton. But *Samson Ago-*

nistes is a deeply ambiguous work, and never more so than in the overdetermined status of Samson's "rouzing motions," somatic mysteries at once absolutely central and resolutely encrypted. If we are, so to speak, to un-Knott the tangles that coil around these inward motions, what might emerge within Samson's final act of melancholic and masochistic martyrdom is precisely the possibility that Fish and Knott find so implausible and so unwelcome within Milton's late poems: perversion. That is, Knott's alternatives—*either* Samson's final act is a display of divine power *or* Samson is ambiguous and perverse—might themselves be a symptomatic version of Samson's own, prior false opposition: *Either* this day will be the day of my great act *or* it will be my last day. In fact, of course, it is both, and the same could be said for the choice between religious faith and literary ambiguity. For Milton, Samson's final act is simultaneously divine and perverse, martyrdom and suicide. Despite Fish's pronouncements to the contrary, there are conflicts upon the page in Milton which he is not able to theologically frame his way out of, and melancholy is just such a case. This is not because melancholy is not an example of "error" but precisely because its conceptual incoherence registers a specific intellectual history that sidesteps providential rescue. It was already an aporetic chasm before Christ was born, and in locating its corrosive classical pagan presence within his Hebraic strongman, Milton compounds two separate cultural distances from the difference-collapsing reach of Fish's critico-Calivinist framework. As Gordon Teskey has stated, "the irony of *Samson Agonistes* is so pervasive, and so uncontrollable, that it reaches out of the text to seize us, ironizing the process of reading itself."[34] Melancholy congeals this all-pervasive irony into an inescapable yet unverifiable local form. First, melancholy can be understood literally as black bile, the liquid substance that further degrades Samson's already-emasculated and weakened self, constituting both a shaming consequence of failure and a redoubling spur to further shame. But in the broader historical scheme of a transvaluative ethics of typological heroism in which suffering and passivity will come to overpower power itself, melancholy also constitutes the sorrowful condition of divine self-sacrifice which heralds a new heroic mode, and hence a new masculinity founded upon the melancholic mourning of the lost object of the heroic and violent "old" masculinity: Whether in the garden at Gethsemane or looking down upon his torturers, Christ as the "man of sorrows" models a kind of passionate, passive masculinity that tantalizingly borders upon melancholy. But melancholy cannot be securely located on either side of this division (cause or effect? curse or gift?) because it cannot be securely located on either side of the differential force field put in place by its competing Galenic and Aristotelian definitions.

In his *Sketch for a Theory of the Emotions*, Sartre describes emotion as a kind of a strategic gesture of problem solving in the face of a seemingly insoluble dilemma posed by a resistant world: When the rational world of utilizable tools breaks down, a world of emotional "magic" takes its place.[35] Early modern authors intuitively grasped the function of drama as a virtual space in which to represent, and morally evaluate, cruel and sudden acts as such limit cases of acceptable problem solving; Webster's oft-quoted lines from *The Devil's Law Case* put it thus: "For sinne and shame are ever tyed together / With Gordian knots, of such a strong thread spun, / They cannot without violence be undone" (II, iv, 48–50). In the case of Samson, melancholy summons a diagnostic epistemology of outward signs and inward feelings that stands in a complex relationship of mimicry and parody with the religious mystery of divine inspiration and sacred violence at the core of the Samson story: In a bitterly ironic fusion of public and private reckoning, by the end of the text, outward theaters of display and spectacle and inward theaters of conscience and self-scrutiny are not only metaphorically but literally collapsed onto each other. Yet the apparent function of this closure is not the production of melancholy as an affective experience for the audience. Rather, in the reportage of the Nuncius and the elegies of the chorus and Manoa, the text's closure aspires toward melancholy's properly tragic evocation and purgation. Milton often concludes his late works with equivocal grace notes that flag, and then dispel, the claim of sad passions: consider the invocation and cancellation of pathos in "Some natural tears they dropped, but wiped them soon" that closes *Paradise Lost* with the partial erasure of sorrow.[36] *Samson Agonistes* is no exception, and the text famously closes with the declaration that at long last those who have witnessed Samson's legacy have achieved "calm of mind, all passion spent." I hope to have shown that at least one other candidate for the passion which has spent itself in a final aporetic act of violence, the passion whose purgation has at long last soothed an anguished and self-critical mind, is neither the love of God nor the hatred of the foe, but only melancholy.

Yet nothing has been brought to an end in marking out this final resting place, nor could it be. In the challenges to interpretation posed by his self-encrypted interiority both within and without Milton's text, the mysterious overdetermination of explanations for Samson's encrypted "rouzing motions" exemplifies the social dynamic of melancholy affect as an endlessly generative problem for the assemblage of audiences and spectators gathered in its wake. Fittingly, it concludes my account of this vexed dynamic within early modern English culture on a note of chronic recalcitrance rather than cure. The only cure is death.

Epilogue *Disassembling Melancholy*

Melancholy never stops, but this book must. Before it does, I wish to both recapitulate and extend my argument, placing a final pressure upon the question of melancholy as "matter," in order to think about how and why melancholy still matters, and where it continues to circulate, today. At the risk of droning melancholically on, I hope to do four things: First, I hope to simply rehearse the multiple definitions of the phrase "melancholy assemblage" in play throughout this book. Second, I hope to locate melancholy's curiously polychronic position within neuro-reductionist research into the brain science of emotion. Third, I want to sharpen the profile of melancholy as an epistemologically troublesome kind of matter by placing it against the hostile ontological backdrop of recent work in "object oriented" philosophy. Finally, I want to track the material poetics of melancholy assemblages at the present time through readings of two particularly expressive examples drawn from music and film: the "depressive black metal" of Striborg and Lars Von Trier's film *Melancholia* (2011). In both Striborg's music and von Trier's film, melancholy as a personal emotional stance is connected to the materiality of "earth" as elemental, environmental, and planetary ground, offering up case studies in how the interpretive/affective knot of the melancholy assemblage carries forward and outward into the present. It is my hope that in the case of all these final gestures, the uncanny persistence, generativity, and recalcitrant critical force of the melancholy assemblage emerges anew, modeling in different ways not only what kinds of assemblages melancholy can bind together, but also how the intractably variable limits of what we can know about melancholy as physical and discursive "matter" are themselves integral to that polymorphous continuity.

The titular phrase "melancholy assemblage" has resurfaced throughout the preceding chapters of this book, with the term risking a certain elasticity of application as it has expanded or contracted across scalar registers. At its close, permit me to revisit the polysemous field of overlapping senses accumulated so far. In this book, I have variously defined a melancholy assemblage as:

1. The melancholy body arranged in space as an iconic, instantly legible object for interpretation. The body itself, particularly when organized into the self-propping stance that links Dürer to Hilliard to Oliver to Ader, constitutes a basic melancholy assemblage, but this basic object can then be looped in a modular manner into relationship with the grounding earth upon which it is posed.

2. The social network defined by a scene of symptomatic presentation, in which a potentially melancholic body is made available for the diagnostic labors of others through solicitation, display, or self-disclosure. From Armado to Biron to Antonio to Hamlet to Samson, the melancholy assemblage expands outward from the scene of presentation to include those nearby who spectate and speculate upon the interiority of an allegedly melancholic body. This expansion process includes the viewers of paintings and plays and the readers of texts within the radius of the assemblage. From a psychoanalytic perspective, "scene" can also be translated as "symptom" and "fantasy."

3. A text composed of found fragments whose organization generates an affective intensity that registers loss, despair, or anxiety. In the case of both Burton and Benjamin, the solicitation and frustration of interpretative energy becomes a defining rhythm or grammar of presentation through which their texts manifest a melancholic form of negative authority. The experience of reading constitutes a scene of diagnosis in relation to an implied authorial presence.

4. A community of witnesses, mourners, spectators, or readers who are constituted through the identificatory legacy generated by a melancholic person or text, sharing in what I have termed its perceptual community. This community is constituted by the mingling of affective flows and epistemological speculations, producing both investment and skepticism in its percipients. Whether we are mourning Hamlet or Samson or simply looking at an emotional body, the burden of our own coparticipation in the feelings and knowledge of others constitutes a social assemblage that includes us in what we see, read, and experience.

At varying levels of organization, the phrase "melancholy assemblage" can thus denote a body, a symptom, a fantasy, a scene, a text, and a community, and the total set of relations possible across and between these distinct levels. The inclusion of some definitions inside of others is intentional and draws inspiration from the complexity theory of Manuel De Landa, which seeks to nest accounts of complex political formations such as nation states and cities within an inclusive and scalar ontology.[1] For my purposes, assemblage theory simply best suits the promiscuous range of examples within the intellectual historical archive of melancholy that I have attempted to describe and unpack because it thinks identity as expressive consistency without essence. Departing from De Landa and from the antiepistemological tenor of most post-Deleuzian philosophies, I have insisted upon the epistemological (or simply interpretive) dimension in play within that archive. I do not aspire to systematicity, but it is my modest hope that the resulting theory of the melancholy assemblage defines a variable but expressively consistent range of phenomena. Such a theory might constitute what Eve Kosofsky Sedgwick, herself an enviably flexible theorist of epistemology's appeal and limits, once termed a "project of nonce taxonomy": a categorical imagining that redescribes and cocreates as it gathers.[2]

To return to a metaphor introduced at the beginning of my argument, the melancholy assemblage is thus not a unified system but an ongoing "problem space" of cumulatively overstuffed traditions, relentlessly overproductive aesthetic conditions, incomplete voids, and inviting openings in social knowledge, contoured by discourses that worry at their own edges and in the process rearticulate themselves into new subassemblages with an open-ended set of potential constituents and components. The term *problem space* comes from Daniel Dennett's description of AI researchers' attempts to infold contradiction-resolution into their working models of the parallel processing architecture of the mind:

> Impasses are basic building opportunities in the system. Conflicts are not automatically dealt with by a presciently fixed set of conflict-resolution principles (an authoritative traffic-cop homunculus already in place) but rather are dealt with nonautomatically. An impasse creates a *new* "problem space" (a sort of topical workspace) in which the problem to be solved is precisely the impasse. This may generate yet another, meta-meta

> traffic problem space, and so on—potentially forever. But in practice (at least in the domains modeled to date) after stacking up problem spaces several layers deep, the topmost problem finds a resolution, which quickly resolves the next problem down, and so forth, dissolving the ominous proliferation of spaces after making a nontrivial exploration through the logical space of possibilities.[3]

Dennett's narrative of problem solving provides a handy sketch of the historical trajectory of melancholy as an epistemologically intransigent object whose very intransigence proved perversely productive of the assemblages of artistic expression and social connectivity that make up this book. The impasse between Galen and Aristotle did not shut the system of humoral thought and explanation itself down, but it did not resolve that system into a centrifugally opposed set of precisely two rival camps either, with clearly marked Galenists on one side and committed Aristotelians on the other. Rather, melancholy became a discursive site of conflict, and distinct agents produced discreet responses to the impasse as they thought the humoral body through, with each melancholy portrait, poem, dramatic character, and diagnostic decision further extending and altering what could count as an example of melancholy, a case. In this manner, over several centuries, the logical space of possibilities defined by the wide disjunction between pathological and genial melancholy, and the wider tetrad of the four humors as a system, was explored and contoured. As the figure of the melancholic became cumulatively richer in rival explanations and significances, this problem space of melancholy became "stacked several layers deep" with subproblems, local variants and new complications. When medical science changed, the rival theories of the humors that defined the shape of that problem space were parted out and replaced. The impasse between Galen and Aristotle was never "solved" because it could not be solved within the terms that produced the conflict; rather, it was inhabited until the premise of humoral materialism upon which the impasse rested was itself abandoned. Eventually, the melancholy assemblage as Oliver and Shakespeare and Burton and Milton variously understood it was disassembled and reintegrated into new formations. But melancholy as a term for a mood or feeling of alienation, isolation, and mysterious, unsatisfiable longing has yielded ground only tenuously to the pharmaceutical and diagnostic supremacy of "depression."[4]

Gearshifting into the depressive present, a blackened spectre still haunts the contemporary science of emotion: the spectre of early modern melancholy. A case of such polychronic haunting occurs in *Looking for Spinoza*, a recent text by neuroscientist and popular author Antonio Damasio. Not content with honoring Baruch Spinoza for anticipating his own field in his title, Damasio goes on to cite an unexpected early modern medical authority on the evolutionary basis of feeling: William Shakespeare. Beneath the insistent banner of the subheading "Trust Shakespeare," Damasio writes "Trust Shakespeare to have been there before," securing Shakespearean provenance for his own field of inquiry.[5] He then engages in a burst of close reading, explicating the passionate outburst in *Richard II* in which the despairing, deposed monarch Richard calls for and shatters a mirror in front of his ascendant rival Bolingbroke as a scene in which "in just four lines of verse, Shakespeare announces that the unified and apparently singular process of affect, which we often designate casually and indifferently as emotion *or* feeling, can be analyzed in parts."[6] By way of context, here is the full scene in question:

RICHARD: A brittle glory shineth in this face,
As brittle as the glory is the face,
He shatters the glass.
For there it is, cracked in an hundred shivers.
Mark, silent King, the moral of this sport:
How soon my sorrow hath destroyed my face.
BOLINGBROKE: The shadow of your sorrow hath destroyed
The shadow of your grief.
RICHARD: Say that again:
"The shadow of my sorrow"—ha, let's see.
'Tis very true: My grief lies all within,
And these external manner of laments
Are merely shadows to the unseen grief
That swells with silence in the tortured soul.
There lies the substance, and I thank thee, King,
For thy great bounty that not only giv'st
Me cause to wail, but teachest me the way
How to lament the cause. (*Richard II*, 4.1.277–292)

In separating internal "grief" from external "laments," Damasio argues that Shakespeare has named into existence a fundamental conceptual distinction between public emotions and private feelings. Like Richard shattering the glass, Shakespeare shatters affect itself into its constituent parts, creating a dramatic taxonomy that anticipates the complex articulation of these components of bodily experience within contemporary neuroscience. When scientists debate the status of claims about the connections between privately experienced psychological feelings on the one hand and quantifiable neurochemistry on the other, they are building upon Shakespeare's account of affect, inheriting his distinction between what we feel within ourselves and how it shows up as an object of knowledge available to others. Shakespeare has been here before, and in studying the phenomena of affect we are returning to our collective early modern origin, which uncannily anticipates contemporary brain science.

Almost. For at this point, Damasio confesses that, however exactly right the Bard was to have sensed that affect is internally divided, Shakespeare got the relationship between emotional "shadow" and felt "substance" exactly wrong. Instead of public emotions arising as the causally and temporally secondary "shadow" of prior private feelings, the facts suggest that the opposite is the case. By the time these inner events show up for us as personal, psychologically acute feelings they have in a sense already "happened." For they have already been registered as medically measurable emotions (and measurable means "public" in the technologically and epistemologically available sense that matters to science): variations in heart rate, blood pressure, galvanic skin resistance, and brain activity. Tellingly, Damasio takes Shakespeare's failure to predict modern neuroscience as his cue to revise the very speech he has quoted and praised: "Here is what Richard *should have said*, in effect, with due apologies to Shakespeare: 'Oh, how this external manner of laments casts an intolerable and unseen shadow of grief in the silence of my tortured soul.'"[7] Rewriting Shakespeare to fit current concerns, our scientist enters the distinguished company of Hollywood screenwriters whose dialogue credits have saddled the plays with "written by William Shakespeare with additional dialogue by Sam Taylor." Getting there first but getting it wrong, Shakespeare is someone we have been ordered to trust and warned away from trusting.

Simultaneously reassuring and disturbing, Shakespeare is someone whose immortal words we feel compelled to cite and repeat—but also to correct and transform.

I dwell upon this affectionately patronizing run-in with the Bard not because I intend to conclude this book with a historicist jeremiad against neuroscience but because this encounter between Renaissance art and the modern science of feeling provides a convenient allegory for the ambivalent terms of survival offered to the past by the present. Damasio's gesture is a common one that summarizes our ambivalent inheritance of early modern culture, in which we are stuck in an asymmetric relationship of debt and antagonism toward undead ancestors that we have been taught to love but cannot actually accept. Nor are the humanities immune to this predicament, for the curious dynamic at work in this entanglement with Shakespearean meaning is no less present in the vexed questions that arise in literary scholarship when we treat Shakespeare as a lens through which to confront our own political, religious, sexual, and racial contemporaneity. The urge to insert a "here is what Shakespeare should have said" is not Damasio's alone.

What remains sticky and unfinished in this encounter is not only the value but the meaning of Shakespeare's own text and with it, the material status of melancholy. For before there were fMRIs and CATscans, there was an early modern physiology that would have supplied entirely different significance to the lines that Damasio hastily rewrites. In referring to an inward grief and stating "There lies the substance," Richard indexes not only an evanescent psychological phenomena to be set in opposition to external laments but also a literally substantial and materialist physical stuff: The substance that lies within an inward "there" is the black bile of melancholy. That is, sorrow was understood to have a robustly physical basis, and in referring to an inner substance Richard is indicating a cause that, in its own intellectual historical moment, would have been regarded as just as scientifically valid as Damasio's own clinical procedures are today. By which I mean, of course, quite obvious to some and not at all clear to others. As I hope to have shown in this book, theories of early modern affect were just as debatable, just as "shattered" into fractious and contrary positions, as the plural worlds of contemporary science and critical theory are today.

This allegory illustrates more than the cultural capital of prestigious names, and reaches into the difficult question of how our activities of reading and interpreting early modern texts might anachronize the present moment. The problem of melancholy substance offers an instructive precursor to current conversations and disputes by lodging an "unselfsameness" within the cultural/scientific scene of affective interpretation: what do we think we are explaining when we make material assertions about mirror neurons or correlate talk about our feelings with the empirical evidence of localized brain function? Do we really need empirical justification in order to make claims about the feelings and emotions we struggle to denote and comprehend?[8] Insofar as some of our most scientifically sophisticated contemporary models for how to think about the materiality of affect are still engaged in methodological dialogue with Shakespeare, Descartes, and Spinoza, it can be argued that early modern affect has provided both the most canonically exemplary case studies of emotional expression and the most ambitious philosophical accounts of how to think about the emotional body. Insofar as such encounters between modern readers and early modern texts are riven by a need to dematerialize early modern physiology in order to render those texts sufficiently pliant to rhetorical reuse, they betray the very culture they would seem to extend and celebrate.

It would seem only natural to respond in turn by rigorously rematerializing early modern affect theory, taking it back to the pre-Cartesian Eden of a lost emotional ontology. But the force of my argument in this book is that it is not sufficient simply to reverse the polarity, though many critics—and not only new historicists and cultural materialists—have done so. Thinking the "thingliness" of melancholy cannot stop at the moment of reasserting its status as an article of faith for adherents to a clinically materialist Galenic doctrine, not only because that doctrine is insufficient as an explanation for the curious persistence and variance of melancholy as an aesthetic category but also because such explanations fail to credit the extent of early modern skepticism occasioned by the performative dimension of humoral temperament itself within the very moment of its cultural circulation. Melancholy was never reliable, never trustworthy.

What is needed is not only a preliminary materialist turn but also a secondary epistemological turn in which the consequences of melancholy's hybrid status between substance and sham are experienced as a problem for knowledge, a jarring of received wisdoms which had distinct outcomes within philosophy, drama, politics, urban experience, and other domains. It is my hope that this book has modeled what the concerns and consequences of such an epistemological turn might be. The conceptual unity of melancholy as a singular object of knowledge must be abandoned in favor of the dispersed and heterogeneous complexity of the assemblage, a manifold that distorts historical experience and troubles the prospect of intersubjective knowing both within the period and across the division between then and now. Accordingly, our stance toward early modern texts on melancholy should not be one of warmth and "trust," but care and caution.

Damasio's encounter with Shakespeare flags melancholy's questionable ontological status in a manner which can help us think about the place of discredited ideas within a social ontology and in the process, suggests that emergent forms of "new materialism" have unfinished business with the epistemological monsters still prowling in their midst. Where might we locate melancholy assemblages built out of purely hypothetical liquid juices within a material world of substance? To answer this question, let us start with realist ontology at its most basic level, where "matter" names the everyday world of extended things, substances, and physical stuff. This is, at least, the first and most common sense definition of "matter" that comes to mind: the billiard ball world of mechanistic action, the street on which one kicks a stone with a hearty anti-idealist cry of "I refute it thus." Complicating things, let us grant that "matter" also names a polysemous concentric overlap between thingly, substantial entities and discursive entities whose inscription and circulation in language and thought is itself necessarily expressed materially in the world (even the words "round square" exist as words extended in space on this page right now). Neuroreductionism promises an increasingly detailed account of that material basis for thought, memory, and emotion, and this includes "faked" emotion and thoughts and speeches about unreal things, entities once confined to the ontological quarantine of "Meinong's jungle."[9] In this expanded sense of matter as expression, anything that can be a kind of subject matter is therefore no less a kind of matter than Styrofoam.

For some this leap is a theoretical given, and for some it represents an appallingly elastic sort of sophistry about the existence of objects. If we pick up the issue from the handle of "existence" rather than from the handle of "matter," we can bear in mind that the capacious framework of social ontology grants existence to money, marriage, promises, myths, and lies no less than to butterflies and slave ships (try saying "promises do not exist" to someone the next time you break one).[10] Substance becomes pervasive. A fast-rewind and fast-forward philosophical montage suggests that perhaps it always was. The early modern dawning of the Cartesian distinction between *res extensa* and *res cogitans* itself constitutes a story about two kinds of substance: Both are ways of thinking through what the term *substance* might conceptually and/or materially capture, and it is only in the wake of subsequent philosophical developments that "immaterial substance" became increasingly oxymoronic.[11] For Descartes and his early modern critics, the distinction births connectionist and correlationist nightmares whose intractability drives him to the miraculous pineal solutions offered in *The Passions of the Soul*, the infelicity of which, in turn, prompts Spinoza's radically preemptive counter-assertion of the fundamental unity of all substance in the *Ethics*: "The order and connection of ideas is the same as the order and connection of things."[12] Fast-forwarding further: In the wake of Cartesian dualism and Kant's critical turn, various proponents of object-oriented philosophy and social ontology seal the resulting chasm through a wide range of distinct but overlapping ontologies which revisit the Spinozan moment of unity but assert new elements and components: becoming and immanence (Deleuze); actors, networks, and actants (Latour); assemblages and emergence (De Landa); and objects (Harman, Bryant, Bogost). These various figures do not necessarily share ontologies, but they share a hostility to Cartesian and Kantian sundering of minds from world and attempt in different ways to work around phenomenological closures and finitudes.[13] In many cases, this urge to achieve escape velocity from the old, divided world is combined with a materialist turn toward a realist ontology of thingly substance that is reaffirmed in its place, and such moves tend to situate what one might call purely discursive objects comfortably within a broader pantheon of objects. Graham Harman's declaration that, qua object, "Reality does not matter: mountains are no more objects than hallucinated mountains" offers a portable case in point.[14] We could call this move, and moves like it, "concentric," insofar as these object-oriented ontologies tend to subsume discourse-matter within thing-matter under an overarching umbrella of matter-as-expression.

And yet this concentric overlap cannot entirely rescue us from the traction of the naïve topographical intuition that there must be outlying margins and a central void at the edges and center of these rings of material belonging: a gap, an empty space, of that which is not matter on the "outside" of materiality: the immaterial, the empty set, the void. This *terra incognita* would define what remains outside the reach of these concentric circles of thingly matter and discursive matter. And, correspondingly, if "subject matter" counts as matter, that there must be a central void at, as it were, the inner border of what has not yet emerged as a discursively available kind of subject matter: the acoustic unsaid, the cognitive unthought, the linguistically nameless, the temporally virtual, and the psychically unconscious.[15] There are means of approach in both directions. To engage the ontological status of that outer void is to do metaphysics; to index the causal effectivity of that inner void constitutes the simultaneously epistemological and psychoanalytic counter-supplement to the monist turns of the so-called "new materialisms." From conceptual and discursive entities to the heels of our shoes, the universe is infinitely and completely expressive as material—except where it is not.

To return from the macrocosmic wide view to the microcosmic terrain of my particular argument in this book: *Melancholy names therefore a historically specific early modern epistemology effect within and upon what can show up affectively as matter.* Humoralism projected the subject matter of melancholy as a received classical theory onto the physical surround of the material, substantial world of bodies and symptoms through ongoing assemblages of ascription, projection, and diagnostic "recognition." Subject matter became a posited form of substance, which in a direct sense never appeared and yet which never failed to appear. That is, since the material support, black bile, "did not exist," there cannot have been a thingly substance for early modern people to feel within themselves or see at work within their bodies. And yet, from the perspective of a social ontology, it is, of course, entirely the case that melancholy "existed" as an object of knowledge, insofar as countless early modern men and women recognized themselves as melancholy, saw the melancholy meanings implicit in each other's behavior, heard melancholy as both tone of voice and mannerism.

Melancholy thus constitutes an ideal means through which to qualify, complicate, and partially revise where we take the boundaries of early modern material knowledge to fall. Though its archive has an expressive consistency, melancholy constitutes a "fugitive matter," which cannot be fixed at a singular, particular point on the map of social ontology or intellectual history: For some, it is a medical fact of life, the liquid essence of embodied temperament. For others, it is fad and a fraud, an urban fashion to be assumed and discarded at will. But on either side of thingly materialism and the discursive realm of social ontology, the black spring of melancholy never dried up, but kept flowing and circulating as a stream of assemblages tethering bodies and signs together across space and in time, subject to deformation and unexpected revivals. Melancholy continues to circulate and recombine at the present moment: immortal, transpersonal, rampant.

Where can melancholy be found? A better question: Where can it not? At once chronic and quotidian, the ubiquity of this emotional orientation inherits its already-elastic early modern symptomatology and pulls the concept further apart in contrary directions. Still swinging between genial and pathological poles long after the demise of humoral medicine, "melancholy" now defines both an (over)familiar pleasure and an uncanny danger. Today, the word can index a world-weary posture of quiet resignation—experienced as a not-altogether-unpleasant stance of sorrowful regard for the world's transience—but it can also show up as a crippling downward slide toward the lowest depths of alienation, self-hatred, and suicidal despair. Just as the term was both celebrated by Florentine humanists and mocked on the public stage by London clowns in its historical moment as Renaissance epidemic, so too at present there are "highbrow," "middlebrow," and "lowbrow" melancholic variants, rival strains which circulate among distinct coteries and constituencies, where they are evaluated, consumed, or endured with wildly variable stances of sympathy and suspicion. Now as then, melancholy is up for grabs: academic fetish, garish commodity, subcultural badge, satiric target.

At the close of this book, I do not have the space required to track melancholy's extraordinarily insistent survival into the present time within the overlapping domains of art, music, film, fashion, and popular psychology; there are far too many examples of contemporary melancholy assemblages to choose from. In order to provisionally sketch the sheer bandwidth of melancholy's reach across global culture, I have isolated two contemporary examples in different mediums from different parts of the planet: the work of the "depressive black metal" musician Striborg and Lars von

Trier's film *Melancholia* (2011). Alternately marginal and high-profile, aimed at cult and mainstream audiences, reaching from their respective origins in Tasmania and Denmark toward a "flat" global marketplace, these two examples offer case studies in the material poetics of melancholy modernity, mutating representational strategies inherited from the intellectual archive of early modern black bile while extending them into the present.

Striborg is the underground musical solo project of an Australian individual who identifies himself only by the name Sin Nanna, helpfully glossed by the online *Encyclopedia Metallum* as "the name of a Sumerian god of the crescent moon."[16] Active since 1997, Sin Nanna currently resides in Hobart, Tasmania, where he records and releases LPs, CDs, and DVDs to an avid global fan base of devotees of "depressive black metal," a style of extreme heavy metal centered upon lyrical themes of ennui, abjection, and mourning.[17] Epitomizing this rapidly expanding microgenre, Sin Nanna's catalogue of Striborg titles are so relentlessly transfixed upon the twin poles of despair and hatred as to court accusations of self-parody. Consider the following litany of album titles that stretches from 1997 to 2011: *Cold Winter Moon*; *Misanthropic Isolation*; *Spiritual Catharsis*; *Black Desolate Winter / Depressive Hibernation*; *Embittered Darkness*; *A Tragic Journey Towards the Light*; *Ghostwoodlands*; *Solitude*; *Journey of a Misanthrope*; *Autumnal Melancholy*; *Foreboding Silence*; *Perceiving the World With Hate*; *Roaming the Misanthropic Forests*; *Black Hatred in A Ghostly Corner*.[18] Striborg's song titles further articulate this obsessive cultivation of negative affective stances, crucially articulating his vision through the invocation of a "blackness" that is imagined as equally aesthetic, magical, natural, seasonal, and emotional: "The Manifestation of Black Residual Energy," "Black Animism," "Entangled in the Black Roots of the Forest," "A Lonely Walk in a Desolate Cold Pine Forest," "Meandering in Sorrow." One could go on.

To put it mildly, this is not a musical subculture that asks for, or rewards, "close reading." Insofar as the deeply repetitive musical structures and unintelligible vocal delivery blur its expressive urgency into a seemingly endless, static experience, the work of depressive black metal artists such as Striborg repels such interpretive practices. Indeed, as a genre black metal offers a resistance to identification through a militant distortion of meaning at every level: Fast-picked guitar riffs are so overblown that they cannot be parsed except as buzzing drones; words are gargled and screeched so that they cannot be made out; murky, deliberately lo-fidelity recording values turn individual drum hits into a nonstop puttering sound; thorny graphic design renders logos into illegible ciphers that can only be decrypted by cognoscenti, and so forth. Against the backdrop of this already-willfully hostile and inward-looking subculture, the particular microgenre of specifically "depressive" black metal turns away from the warlike and aggressive stances typical of more sanguine forms of heavy metal and black metal through a specific lyrical and aesthetic fixation upon sorrow, loss, and self-killing.[19] As I have written elsewhere:

> The circulation of black bile within black metal, aptly described by Nicola Masciandaro as the "thrown conceptual space" of Melancholy Black Metal,[20] models with particular virulence a kind of temporal distortion already implicit in the early modern archive of melancholy representation: the past of an archaic origin and the endpoint of an extinct finitude yet to come are both brought into a present moment.[21]

Transposing humoral aesthetics from Tudor drama to the lyrics and artwork of a burgeoning global subculture, depressive black metal opens a wormhole within the mosh pits and chat rooms of modernity that tunnels backward through early modern intellectual history toward ancient Greek medicine.

The lyrics to "Digging a Ditch to Die In," from *Embittered Darkness / Isle De Morts* (2006) offer a particularly telling example of Striborg's contemporary reinscription of black bile. The song's "black veins of melancholy" connote both natural cycles of death and decay and an abject personal impasse of frustration, boredom, and despair:

> *Digging a Ditch to Die In*
>
> Sometimes I am so fed up with my own existence
> Times are long gone, time is quickening and so short
> Black veins of melancholy support my life system
> My life among the trees of sadness and woe
> Too much reality becomes overwhelming in this dull existence
> I only long for the dark and mysterious ways
> They are always there with an open mind (drugs help)
> The nightmares reveal the real world to me

The world of nightmares is my real existence
Life long pressures and sickness in the inevitable end
I sometimes wish I could hide in the ground for eternity
That day will become eminent anyway
But sometimes I really could dig my own fucking
Ditch to die in, starving myself to death
At least I would be out harm's way—humanity
That way I don't need to see your ugly face again
You won't need to see mine I am just as insignificant
As the pile of dirt I just dug up!
So why don't you join me in ending your worthless life
And fertilize Mother earth's rapacious ground![22]

At once ludicrous and frightening, the stoner humor of Striborg's parenthetical admission that "(drugs help)" will undoubtedly raise a laugh of recognition from some readers, and a snort of disdain from others. I am not interested in pedantically detailing the song's infelicities of usage and grammar, but in hearkening, all chuckles aside, to the song's argumentative arc. We move from the polychronic pileup of melancholy temporality as simultaneously "long gone," "quickening," and "short" to a tenuous evocation of a sympathetic connection between the "black veins of melancholy" inside the body and the natural surroundings of the trees. But once this possible stance of ecological affiliation with nature and against humanity is imagined, it comes undone, and we conclude with the song's final invitation to the listener to join Sin Nanna in suicidal death. In a self-differential conflict which may seem familiar at the end of a melancholic gallery that includes Shakespeare's Hamlet and Milton's Samson, here too time runs fast and slow, emotional valences are alternately sharp and dull, and there is a paradoxical mixture of intensity and malaise as the song walks a tightrope between feeling suicidal and feeling murderous. Lyrically and sonically, Striborg's "Digging A Ditch to Die In" manages to be both utterly passive and violently berserk at the same time.

Drawing upon melancholy traditions that precede it by centuries, "Digging a Ditch to Die In" thus exemplifies an aesthetic-affective position that Scott Wilson has dubbed "melancology," a melancholic ecology in which black metal aesthetics constitute a "negative form of environmental writing" by evoking a natural world that is imagined as ultimately and entirely hostile to life itself; as Wilson puts it: "The plane of immanence of melancology is extinction and nonbeing. All things are destined for extinction; immanent to all being is the irreducible fact of its total negation without reserve or remainder."[23] In a feedback loop already implicit in medieval and early modern melancholy humoral portraiture, earth as the element of melancholy constitutes a dark and cold material base that summons and confirms the interior emotional storms of the melancholic individual, resituating them within an environment whose fundamental orientation toward finitude, stillness, and death thus blurs the line between bodies, things, elements, emotions, and matter within an imagined future scenario of ultimate extinction. As such, Striborg's counter-anthem offers a kind of blackened parodic supersession of the "dark ecology" called for by Timothy Morton in his influential text *Ecology Without Nature: Rethinking Environmental Aesthetics* (2007).[24] Mother Earth is invoked, but not as a maternal point of origin now imagined to be in peril, or in need of some kind of ethical stewardship or political protection. Resolutely negative, the song's narrator speaks on behalf of no positive environmental program other than a dark form of self-sacrificial "fertilization." Mother Earth is the goddess of death, but she is not celebrated as such in a transcendental or Romantic register. Rather, the ditch constitutes a hole of withdrawal, an insignificant pile of dirt. So constituted, Sin Nanna's lyrical gravesite rethinks the gashed black spring over which Oliver's Edward Herbert presided in his Tudor miniature portrait: what was a moment of "Magia Sympathia" between the propped melancholic body and its telluric ground is now rigorously rematerialized as a one-size-fits-all grave. Citing a "long gone" temporality that hearkens to the archive of black bile, the voice of the "black veins of melancholy" within the self call us to die, and to encourage others to die with us, so as to fashion a final melancholy assemblage in which our corpses rejoin the earth.

Reaching the same relentlessly morbid and aggressive conclusion from a distinctly different cultural location, Lars von Trier's film *Melancholia* (2011) offers up a masterfully suffocating traversal of melancholic tropes: in muted blue-black colors, brittle interactions between family members, clients and service providers, bosses and employees collectively limn a social landscape mired in stasis and surplus, yet catalyzed by the anxious prospect of individual and global death. In the center of this field stands the emblematic character of the distracted bride Justine, whose manic glee at her wedding soon cracks open to reveal the frozen rictus within. By the end of the film, she is a walking symptomatic checklist of inhibition, ennui, and despair. Toggling between manic ups and depressive troughs, the film's narrative stalls out in deserts of malaise periodically punctured by mountain peaks of sublime overreach scored to the prelude to Wagner's "Tristan und Isolde."

Hanging perilously above this world of wealthy allegiances gone sour looms the titular Melancholia, a previously unnoticed planet. This impudently antirealist conceit surfaces gradually, first via hearsay and paranoid amateur websites, then as the subject of mingled anxiety and impotent reassurances. In the film's brutal, majestic climax, the ice blue orb is finally disclosed as a remorseless agent of planetary destruction. An alternately too-absent and too-present lodestone which magnetizes the film's characters, this "Melancholia" stands in for Saturn, the cyclically destructive lord of death who comes at the point of apocalyptic closure to harvest humanity and end the world. Tilting the personal against the planetary, the film is thus structured around a basic conceptual division within the intellectual archive of melancholy, which it straddles and renders ultimately coextensive: melancholy/melancholia names *both* a passing mood and a deadly threat.

Initially, the film seems to present us with a lightweight and reductive reading of the emotion for which it is named: The pampered despair of its characters comes on like a gust of transient and baseless affect that will blow over if managed properly (an attitude epitomized by Justine's husband's blasé declaration that "Melancholia is going to pass right by us, and it's going to be the most beautiful sight ever"). Such a construction should be familiar from the opening declarations of Antonio and the yawns of Portia in *The Merchant of Venice*: melancholia as the side effect of economic privilege. Yet this flattening reading is overturned by the actually flattening power of the planet Melancholia itself, which destroys not simply these elite sufferers from stressful jobs and bad marriages but the entire world. The film's ultimate moment of closure as absolute annihilation thus offers a ringing endorsement of depressive realism as the philosophical recognition of being-toward-death on a transindividual, planetary scale. Indeed, the film's ending in a silent blackout at the moment of impact formally enacts this stance in the most phenomenologically direct and violent way possible within film as a medium.

Such is the basic ambivalence von Trier's film puts into place, and it is an ambivalence which should look familiar as a concluding gesture to the archive I have been reading in this book: the both/and of melancholy as empty *and* full, thin *and* deep, false *and* true, private *and* total. Melancholia is both the privileged affective stance of the elite, the side effect of the good life gone bad, and yet, horrifically, a wild guess at a something more that levels economies, species, and world. In the wake of its spoiler-in-slow-motion prelude, the film proper starts as a potshot at the bourgeoisie, but melancholy's broader purchase is revealed by the film's conclusion to transcend its class signifiers and to ratify Robert Burton's definition of melancholy in *The Anatomy of Melancholy* as the "character of mortality": that which emerges triumphant when human beings finally allow themselves to recognize the absolute limit of death and the underlying total indifference of the cosmos.

We start on one side of that spectrum for the first section, "Justine." The stuffy, spoiled lifestyle pornography of the film's opening hour is devoted to the slowly broiling murky interiors and exteriors of a private estate hosting a lavishly wasteful wedding; these scenes rehearse the ready-to-hand cliché that melancholia is the ultimate surplus enjoyed by the rich and idle, the overripe consequence of a plenitude in freefall. The chattering classes are assembled, and their mandate to "be happy" becomes increasingly oppressive and alienating as the mechanical rituals of love in the upper tax brackets fray and drag.

For a long portion of the film, you can sense in the theater the unease of its patrons as a mirroring of what is on the screen. We in the cinema are impressed into service too, for like the gawking wedding guests, we are placed by the handheld, jerky camera a little too close to something intimate that is going awry, very slowly but surely, before our eyes. The fact of Justine's sadness ruins things, but at first it seems paltry, selfish, a symptom of the entitlement that surrounds her rather than a critical alternative to it. She accuses her venal boss of being power hungry, but in the disappointments she casually doles out to her husband and in the rapacious coupling with a random flunky from her company we see that her own eccentric orbit can just as easily scan as a power grab of its own: the power to "make scenes," break promises, and ruin the splendor that surrounds her. Here melancholy feels simply like childish ingratitude.

But the glowering storm of impatience this generates is intentional, for it lets the film set up a relational system of emotional investment strung across the mobile force field between the sisters which redoubles in the second section of the film, subtitled "Claire." Claire, as a married mother who manages and fusses, at first patrols the edges of Justine's vision, hovering and steering her fragile sister toward the show-jump-like hurdles of familial expectation; in her own half of the film, the foil figure who stood in for an ambient critique of melancholia succumbs to that very condition herself, which she experiences as a wrenchingly inescapable experience of panic and horror.

The film has been accused of self-indulgence and bathos, but it comes paradoxically most alive only in its most grimly fatalistic moments, in the final exchanges between Justine and Claire. The very futility of adopting "stances" toward planetary death is the unspoken substance of their final conversation; the stakes are at once utterly weightless and overwhelming. Justine's Gnostic pronouncements of negativity ("life on earth is evil") pass judgment upon the stock world of stuffed shirts and tuxedos from which she has awakened into melancholy as a state of profane illumination. "I know things," she declares, with an assurance that recalls both Marsilio Ficino's evocation of the melancholic sage and Freud's characterization of the melancholiac as the bearer of unbearably unhealthy truths. These lines can be read as von Trier's own not-so-subtle declarations of fealty to a character at the seeming expense of the film they inhabit. They are of a piece with the film's earlier, self-conscious bracketing of alternatives to nihilism: Two hours late to their wedding, we are not treated to a scene in a church with a priest offering any sort of religious transcendence. The film offers no real alternative to Justine's suffocating judgment. In a sense, "life on earth is evil" is to *Melancholia* what "chaos reigns" is to the von Trier film that precedes it, *Antichrist* (2009): the absolute of a total negation that rings out as either cosmic or adolescent depending upon your commitments.

As with *Antichrist*'s exploration of grief, in which a bourgeois married couple is thrust up against the fact of death with conflicting conclusions and symptoms, so here, too, the character of mortality—the reality of a non-negotiable endpoint, to individual life and to planetary life—draws out and centrifugally opposes two characters, yet binds them as they mutually translate and care for each other. Justine enters the high romantic embrace of mortality and the stasis of a life-in-death emblematized in the film's car alarm–like repetition of Wagner's keening "*Liebestod*." Claire, increasingly untethered from her control over the fate of her husband and child by the planetary scope of the disaster, struggles gamely to care, love, and protect her loved ones even as the local questions (to go to the village, to serve pancakes) are nullified by the enormity of what is to come, regardless of these choices. Having been encouraged to side with her and against Justine for the first half of the film, we are drawn with her as she too slips into the alternately intoxicating and horrifying vision of extinction that Justine has already intuited and traversed. Inheriting the strategies of perceptual community that undergird the dramatic phenomenology of "Hamlet," *Melancholia* is a film about the spectatorship of melancholia that enacts, rather exactly, the transferential affective legacy of melancholy as a relational network between and across persons which has been the subject of this book.

A key scene offers a case in point: Claire follows Justine and watches her bathing her nude body in the blue light of the risen planet of Melancholia, absorbing its invisible influences. Kirsten Dunst's body is perched on a riverbank in a distanced medium shot, served up for the viewer like Hedy Lamarr in Gustav Machaty's *Ecstasy* (1933), as Wagner's prelude throbs on. But far from becoming invited to invest ourselves emotionally in Justine, whose frontal gaze is offered in a close-up of her face and breasts, we are pulled ineluctably earthward toward the clothed, surreptitiously viewing figure of Claire. In watching Claire watch Justine, we abide in the space between them, and are encouraged to enter into a chain of affective transfer from Melancholia-the-planet to Justine-the-melancholic to Claire as witness and coparticipant in this melancholy assemblage, a chain that binds persons to one another and to the earth through imminent loss, and which conscripts us as viewers into its field of emotional gravity.

But the film does not make such bindings easy, and von Trier deliberately punctures and violates these assemblages as often as he builds them. The exchange between the two sisters when faced with the undeniable fact of the planet's return epitomizes this affective distance: Keeping the wedding planning charade of a bourgeois art–directed atmosphere going, Claire calls for wine on the terrace and a song. Coiling with hatred, Justine spits back sentiments worthy of a black metal vocalist: "You know what I think of your plan? I think your plan is shit. Let's meet on the toilet." These pendular mood swings toward hostility and aggression are expressed only to run backward into care for others. Faced with the anxiety of Claire's child Leo, Justine relents and affirms a transcendent, protective myth with a clear recognition of its impotence. Leo becomes a sacrificial lamb in whose purity we as viewers are invited to invest. Von Trier's bad faith about his own deployment of the emotionally pornographic assertiveness of Wagner is betrayed in a symptomatic knife-flick of dialogue from Justine, who, when cashing out what sort of playlist best suits the end of the planet, mockingly proposes "Beethoven's 9th Symphony." This offloading of shame at the impotence of high culture in the face of apocalypse is the inner lining of the film's own use of Wagner. But the film needs the Beethoven-and-Wagner comedy as a miniaturized variant of its macro level Earth-and-Melancholia cosmology and its interpersonal grounding in the Justine-versus-Claire narrative.

As above, so below. Melancholy assemblage as an affective community constitutes both the engine of destruction and its only, yet ineffective, rejoinder: The pairing of the sisters stands both for and beneath the pairing of the planets. Each interlocks with the other, for the arc of circling, withdrawal, and collision between the Earth and the "flyby" planet of Melancholia, which Claire witnesses on a website and is prevented from printing out by an ominous power failure, also describes the plot arc of Claire's stance toward Justine: at first too close, then angry and pulled away, and finally all too close yet again, in death. Embrace and planetary collision are joined in the final apocalypse.

The film concludes with the ultimate melancholy assemblage: the "magic cave" of twigs which the melancholic Justine constructs around the last human gathering: a metaphor for both the (im)potent structure of cinema's temporary perceptual communities and for the broader status of mortal humanity thrown upon the impassive ground of a planet which will be their grave.[25] Like Adam and Eve taking up the posture of melancholy in the face of the bitter fact of life's transience, these last participants are assembled into a tenuous, living circuit of regard and care, a transient and rickety shelter built within and against what they already know they are powerless to defer. This offers an object lesson in the socially connective force of negative emotions: Justine as the bearer of melancholic truth is also the architect of a temporary structure binding people to one another and to the environment that tenuously supports them. Premised upon sadness and terror at the fact of death, melancholy assemblages bring people together.

In the face of the frightening fantasies of extinction offered by Striborg and von Trier, the kit bag of critique may not contain sufficiently strong medicine. With its inoculations against illusion at the ready, critique would seem to offer a particularly seductive comfort in the idea that the toxic vapors and hallucinations of early modern intellectual history and its uncanny return within contemporary culture can be dispelled through a swab of skeptical astringent and a quick jab at other people's credulity, romanticism, self-indulgence, or false consciousness. But there is no "end" to melancholy, only a continual mutation, transformation, and working through of its intractable downward pull. Here the advice for how to break up, scatter, and disperse the matter of melancholy within suffering living systems offered by Robert Burton at the conclusion to the *Anatomy of Melancholy* is apt: "Be not solitary, be not idle." Not only is this good practical advice; it also holds as a recipe for the contemporary theorization of melancholy itself. The opposite of solitude would be, of course, connectedness, the imbrication of the self in social assemblages, and the interconnection of melancholy assemblages within a larger network of connections, pathways, collaborations, and links. The opposite of "idleness" would be movement, action, and utility. So, to critically "dematerialize" and "disassemble" contemporary melancholy by way of Burton's advice, we need to give melancholy a shove, patch melancholy into new assemblages, new connections, new movements. Connected to one another in open-ended layouts subject to suspenseful cycles of territorialization and deterritorialization whose outcomes are still uncertain, we are no longer solitary nor idle but busily surrounded by others whose ability to parse, recognize, or refuse to ascribe melancholy to us—in the middle of a mosh pit, in a chat room, in the audience at the movie theatre, in the hallways of a conference—keeps the question "was anyone ever really melancholy?" ever more acute and ever more elusive. Facing down the neuroscience of the present from the perspective of the vanquished humoral science of the past, the epistemological assemblages of melancholy still have an estranging power when brought to bear upon the networks of emotional turbulence, depression, and "panic attacks" that surround us in the present. Taking up fragments from the shattered mirrors of early modern affect within the work of Oliver and Shakespeare and Burton and Milton, we still hope to see ourselves. We can turn and walk away from the panic attack in the hallway. But even if we do, we are nevertheless forced to ask ourselves: What exactly is the matter that we are walking away from? Coiled steadfastly around the impossible certainty of our own deaths, melancholy links together both our temporary postures of sorrowful knowledge and the ongoing assemblages (textual, artistic, and social) that we construct to outlast them.

Notes

Acknowledgements

1. Lyn Hejinian, *The Fatalist* (New York: Omnidawn), 2003.

2. Jorge Luis Borges, "The *Biathanatos*," *Other Inquisitions: 1937–1952*, trans. Ruth Simms (Austin: University of Texas Press, 1964), 89.

Introduction

1. John Ashbery, "Album Leaf," *Some Trees* (New York: Corinth Books), 1970.

2. Such tensions about the adequacy of criteria are acutely present today in debates about the status of diagnostic manuals such as the *Diagnostic and Statistical Manual of Mental Disorders (DSM)*. Consider the aporetic conclusion of Harvard psychologist Richard McNally's *What Is Mental Illness?*: "What counts as a mental disorder depends on shifting cultural, political, and economic values as well as on scientific facts about how our psychology and biology can go wrong, producing suffering and functional impairment in everyday life. We'll never have a clear-cut list of criteria that will enable us to identify all instances of mental disorder and exclude everything else." Richard McNally, "So What is Mental Illness Anyway?," *What Is Mental Illness?* (Cambridge, Mass.: Harvard University Press, 2011), 212. What is left unclear is the relationship between this supposed criterial impossibility and the categorical disjunction between the "soft" (social-scientific) and "hard"(biological) sciences itself; are there shifting values on one side and evolutionarily hard-wired living systems on the other, or a recursive loop in which each half unpredictably catalyzes the other?

3. See Michael Polanyi, *The Tacit Dimension* (Chicago: University of Chicago Press, 2009), 4; for more on the meaning of physiognomic expressions in Polanyi, see page 12.

4. William Shakespeare, *As You Like It*, 4.1.10–18.

5. Douglas Trevor, *The Poetics of Melancholy in Early Modern England* (Cambridge: Cambridge University Press, 2004), 31.

6. William Shakespeare, *The Second Part of King Henry the Fourth*, 3.2.256.
7. Ibid., 5.2.118.
8. Edmund Spenser, *The Faerie Queene*, 5.4.21, 553.
9. *OED*, "Assemble."
10. *OED*, "Assemblage."
11. Deleuze and Guattari, *A Thousand Plateaus: Capitalism and Schizophrenia*, 88.
12. For more on the tensions that would preclude any easy alignment of Deleuze and Foucault, see Wendy Grace "*Faux Amis*: Foucault and Deleuze on Sexuality and Desire," *Critical Inquiry* 36, no. 1 (Autumn 2009): 52–76.
13. Alain Badiou, *Logics of Worlds*, 1.
14. Deleuze, *Two Regimes of Madness*, 177.
15. Ibid., 178–179.
16. For recent inheritors of Deleuze and Guattari primarily working in political theory, "assemblage" permits a recognition of the working power of an abstract system upon and within society without reifying that system into a static fossil and that society into a passive instrument in the process. Manuel De Landa, *A New Philosophy of Society: Assemblage Theory and Social Complexity* (Continuum, 2006), see esp. 28. Conceptual modularity enables a recursive relationship between assemblages. Because assemblage theory can "stack" assemblages within each other, the term can convey both a conceptual form and the plural public within which that form circulates and changes over time, and the corporate body of any particular person, and on down to smaller and smaller descriptive levels. For a balanced appreciation and critique of De Landa's position, see Graham Harman "The Assemblage Theory of Society," *Towards Speculative Realism* (London: Zero Books, 2010), 170–199.
17. Ron Silliman, "For L=A=N=G=U=A=G=E," *The L=A=N=G=U=A=G=E Book*, eds. Bruce Andrews and Charles Bernstein (Carbondale: Southern Illinois University Press, 1984), 16.
18. The words "melancholy assemblage" themselves stand in a curious relation to homosexuality. In Sir Edward Bulwer-Lytton's *The Last Days of Pompeii*, the phrase appears in the midst of a censorious description of the effete Roman courtier Glaucus sizing up gladiatorial manhood. Thrilled and disgusted by the effect of contrast, Lytton's prose lurches into an anachronistic fever dream of panicky recognition:

> So have we seen at this day the beardless flutterers of the saloons of London thronging round the heroes of the Fives-court—so have we seen them admire, and gaze, and calculate a bet—so have we seen them meet together, *in ludicrous yet in melancholy assemblage,* the two extremes of civilized society—the patrons of pleasure and its slaves—vilest of all slaves—at once ferocious and mercenary; male prostitutes, who sell their strength as women their

beauty; beasts in act, but baser than beasts in motive, for the last, at least, do not mangle themselves for money! Sir Edward Bulwer-Lytton, *The Last Days of Pompeii* (Paris: Baudry's European Library, 1839), 84.

Here, the words "melancholy assemblage" constitute a canny form of innuendo through which to designate a queer counter-public. Yet Bulwer-Lytton does not present that community as a unity; the assemblage is rendered melancholy through the sundering effect of an internal difference. What makes this cluster precisely "queer" is that it constellates homosexuality as an economic relationality rather than an outcaste founded upon a shared identitarian essence.

19. Walter L. Strauss, *The Complete Engravings, Etchings & Drypoints of Albrecht Dürer* (New York: Dover Publications, 1972), 170.

20. Erwin Panofsky, "The Meaning of the Engraving B70," *Saturn and Melancholy: Studies in the History of Natural Philosophy, Science and Art* (London: Basic Books, 1964), 403–405.

21. Graham Harman, "Object Oriented Philosophy," *Prince of Networks: Bruno Latour and Metaphysics*, 207.

22. Harman, 35.

23. For a discussion of "flat ontology" in the thought of Bruno Latour, see Graham Harman, "Object Oriented Philosophy," *Prince of Networks: Bruno Latour and Metaphysics*, 207. Such a reading of Deleuze might align even rather different readers of Deleuze, such as Alain Badiou and John Rajchman. See Rajchman's introduction to Gilles Deleuze, *Pure Immanence: Essays on a Life* (New York: Zone Books), 2001, and Alain Badiou, "Which Deleuze?," *Deleuze: The Clamor of Being* (New York: University of Minnesota Press, 1999), 9–19, and the appendix of supporting excerpts from Deleuze's own works assembled therein.

24. Deleuze, *Pure Immanence*, 26.

25. Of the many books on the subject, the following texts constitute the classic examples: Lawrence Babb, *The Elizabethan Malady: A Study of Melancholia in English Literature from 1580 to 1642* (East Lansing: Michigan State University, 1951); Bridget Gellert Lyons, *Voices of Melancholy: Studies in Literary Treatments of Melancholy in Renaissance England* (New York: W. W. Norton, 1971); Dr. Stanley Jackson, M.D., *Melancholia and Depression from Hippocratic Times to Modern Times* (New Haven: Yale University Press, 1986). More recent texts on the subject published during the last decade which have informed my own work include Gail Kern Paster, *Humoring the Body: Emotions and the Shakespearean Stage* (Chicago: University of Chicago Press, 2004); Douglas Trevor, *The Poetics of Melancholy in Early Modern England* (Cambridge: Cambridge University Press, 2004); Jennifer Radden *The Nature of Melancholy: From Aristotle to Kristeva* (Oxford: Oxford University Press, 2002); Adam Kitzes, *The Politics of Melancholy*

from Spenser to Milton (London: Routledge, 2006); Noga Arikha, *Passions and Tempers: A History of the Humours* (New York: Ecco, 2007); Jonathan Flatley, *Affective Mapping: Melancholia and the Politics of Modernism* (Cambridge, Mass.: Harvard University Press, 2008); and Darien Leader, *The New Black: Mourning, Melancholia and Depression* (New York: Graywolf), 2009.

26. William James, "On Some Hegelisms," *The Will to Believe and Other Essays in Popular Philosophy* (Cambridge, Mass.: Harvard University Press, 1979), 221. For more on the nature of the "polychronic" within early modern studies, see Jonathan Gil Harris, *Untimely Matter in the Time of Shakespeare* (Philadelphia: University of Pennsylvania Press, 2009). In a note on Hamlet, Harris notes that

> Hamlet's reflections on what imagination can do to the dust of the past suggests that we need to do more than just understand the polychronicity of matter, or how it is marked by the traces of multiple times. We equally need to theorize matter's multitemporality: that is, the ways in which we physically and imaginatively rework matter to produce diverse organizations of time. (Harris, 13)

Harris' formulation can be usefully applied to melancholy as a kind of permanently polychronic material that lodges a resolutely classical form of substance within the bodies and minds of early modern persons.

27. See M. R. Wright's explanatory introduction to his edition of Empedocles, *Empedocles: The Extant Fragments*, trans. M. R. Wright (New Haven: Yale University Press, 1981), 1–93, see esp. 27.

28. In "On Coming to be and Passing Away," Aristotle critiques Empedocles' scheme for its difficulty in explaining how such changes of state are fundamentally possible. See Aristotle *The Works of Aristotle.* Trans. Sir David Ross. Oxford: The Clarendon Press, 1963. Vol. II, *De Corruptione Et Generatione*, Book I, 1.314.

29. This explanation of material changes of state also models movement and mixture across the boundary that separates humanity from the immortal gods, as we see in the conclusion to one of the longest surviving fragments (47 [35]): "And as they were being mixed, countless types of mortal things poured forth, but many, which strife still restrained from above, stayed unmixed, alternating with those which were combining, for it had not yet perfectly and completely stood out as far as the furthest limits of the circle, but part remained within and part had gone out of the frame. And, in proportion as it continually ran on ahead, a mild, immortal onrush of perfect love was continually pursuing it. Immediately what were formerly accustomed to be immortal became mortal, and formerly unmixed things were in a mixed state, owing to the exchanging of their ways. And as they were being mixed, countless types of mortal things poured forth, fitted with all kinds of forms, a wonder to see." (*Empedocles: The Extant Fragments,* 47[35]). Empedocles' account of Love and Strife as organizing principles of the universe carries over into later Renaissance occult and hermetic

writings, such as Giordano Bruno's *De Vinculis in genere* ("A general account of bonding"), an unfinished 1588 text.

30. An index of Empedocles' liminal position at the border of poetry and philosophy is Aristotle's attempt in the *Poetics* to insist that Empedocles and Homer do not belong in the same category simply because they both write in the same poetic meter. See *The Works of Aristotle*. Trans. Sir David Ross (Oxford: The Clarendon Press, 1963). Vol. XI, *De Poetica*, Book I, 1.1447b.

31. See Ludwig Edelstein, "Hippocratic Prognosis," *Ancient Medicine* (Baltimore: Johns Hopkins University Press), 1967, 65–85, esp. 72. For a nuanced account of how conceptions which modern persons would regard as unscientific appear within medical discourse, see Ludwig Edelstein, "Greek Medicine in Its Relation to Religion and Magic," *Ancient Medicine* (Baltimore: Johns Hopkins University Press, 1967), 205–246.

32. Aristotle. *The Works of Aristotle*. Trans. Sir David Ross (Oxford: The Clarendon Press, 1963). Vol. VII, Problemata, Book XXX.1, 953.

33. It is ironic that, from the perspective of his early modern inheritors, Galen of Pergamon was taken to be an antagonist to Aristotle. This is because his intellectual achievement consists as much in the scholarly preservation and consolidation of ancient Greek sources as it does in innovations and clinical successes. As Luis Garcia-Ballester puts it: "Hellenistic medicine of the second century A.D. is a direct heir to the medical schools in the fifth, fourth and third centuries B.C., and furthermore, was conscious of this dependence. The process of 'returning to the sources' of Greek thought culminated in the second century A.D." (Garcia-Ballester, *Galen and Galenism*, 139). Many of the doctrines which were subsequently regarded as definitively "Galenic," such as, for example, the pneumatic theory (the claim that the body's three principal members are the liver, heart, and brain, each of which is dominated by its own particular *pneuma*, or spirit) have clear precedents in the texts of Hippocrates, Diogenes and Aristotle (Sarton, 51). Galen's respectful distance from his predecessors allowed him to synthesize their work, and alternately refute and defend their positions, in effect consolidating his own position as a canonically authoritative commentator upon all that came before him. This had particularly powerful consequences for the history of melancholy, for *Peri crasion*, his defense of the Hippocratic account of the humors, became, in its 1521 Latin translation by Thomas Linacre titled *De temperamentis et de inaequali intemperie*, one of the most important early modern medical texts on the disease (Sarton, 55). See also his *De naturalibus facultatibus (On the Natural Faculties*, also translated by Linacre) and the *De Locis affectis (On the Affected Parts*).

34. Radden, 68.

35. Marsilio Ficino, *Three Books on Life: A Critical Edition and Translation with Introduction and Notes*. Eds. Carol V. Kaske and John R. Clark (Tempe: Renaissance Society of America, 2002).

36. As Zera Fink notes in her essay "Jaques and the Malcontent Traveler," foreign travel outwardly transformed English gentlemen into visible emblems of melancholy:

> Throughout the sixteenth century and the greater part of the seventeenth, black was the customary color for the dress of Italian gentlemen, and from Italy the vogue was carried to France during the second half of the sixteenth century, when Italian influence was strong and it became fashionable to imitate the dress and manners of the Florentines and Venetians. The effect of this fashion on our traveler was simply that he came back to England dressed from head to foot in black instead of the medley of gaily colored fashions from many different countries which was affected by many foreignized [*sic*] Englishmen apparently at all periods during the reign of Elizabeth. (Zera S. Fink, "Jaques and the Malcontent Traveler," *Philological Quarterly*, 16 [1935]: 237–252, 241).
>
> The effect of such foreign imports upon returning to London can be felt in the following lines from Marston's verse satires:
>
> Look, look, with what a discontented grace
> Bruto the traveler doth sadly pace
> Long Westminster! O civil-seeming shade,
> Mark his sad colours!—how demurely glad!

John Marston, "The Metamorphosis of Pigmalion's Image, and Certain Satyres." *Works*. Vol. 3. (London: John Russell Smith, 1856), 200–237, see esp. 221. Marston's lines alert us to the performative dimension of melancholic costume; it is not enough to simply wear black without the supplement of "sadly" pacing in public. English travelers brought home not simply a costume, but an attitude, a doctrine of inner genius and outward grief, a mode of desolate comportment, a posture.

37. Ben Jonson, *Every Man in His Humour: A Parallel-Text Edition of the 1601 Quarto and the 1616 Folio*, Ed. J. W. Lever (Lincoln: University of Nebraska Press, 1971), 109.

38. Paster, 24.

39. Babb, *The Elizabethan Malady*, 66. This is a rhetorically useful simplification of what was in practice a far denser and more chaotic affair. The transmission of melancholy is marked by a sequence of translations, misattributions, scholastic distinctions, recombinant mutations, and wishful thinking on the part of synthesizing "neoterics," physicians, and humanist philosophers alternately intent on preserving or refashioning the "Galenic" and "Aristotelian" source texts. In the process, these authors reify errors that they retroactively projected into powerfully authoritative positions, constructing and defending doctrines at odds with each other and with the evidence provided by the emergent practices of first hand anatomical observation.

40. Babb, *The Elizabethan Malady*, 180.

41. Paster, 21.

42. The term "problem space" comes from Daniel Dennett's description of AI researchers' attempts to infold contradiction-resolution into their working models of the parallel processing architecture of the mind. Daniel Dennett, *Consciousness Explained* (New York: Back Bay Books, 1992), 297.

43. There is a precedent within analytic philosophy for precisely such a seemingly recursive formulation; as R. M. Sainsbury points out in his essay "Objects without Boundaries" these strategies are particularly useful when attempting to fashion working definitions of so-called "boundaryless" concepts:

> A semantic theory can quite legitimately be homophonic, that is, can reuse in the metalanguage the very expressions whose object-language behaviour it is attempting to characterize. Asked how a boundaryless predicate like "red" works, my first response would be "red" is true of something if that thing is red. [. . .] In fuzzy logic, for example, the distinction will emerge in the structure of J-sets. Those associated with vague predicates will, for some or many objects as arguments, deliver numbers intermediate between 0 and 1 as values; whereas a sharp predicate's J-set will have 0 or 1 as the only values, whatever objects are arguments. In a homophonic theory the information is in a way present, but is inexplicit. R. M. Sainsbury, "Concepts without Boundaries," *Vagueness: A Reader*. Eds. Rosanna Keefe and Peter Smith (Cambridge: MIT Press, 1996), 260–261.

We can regard the total set of melancholy assemblages as a kind of culture-wide "J-set" keyed to a grayscale of values between 0 and 1.

44. For a thoughtful philosophical critique of the "unmasking" hypothesis that drives many cultural studies arguments about the alleged social construction of a given phenomena, see Ian Hacking, *The Social Construction of What?* (Cambridge, Mass.: Harvard University Press, 2000), see esp. chapter 1. "Why Ask What?," 1–36, and "Madness: Biological or Constructed?," 100–125.

45. Galen, "On the Sorites," *Vagueness: A Reader*. Eds. Rosanna Keefe and Peter Smith (Cambridge, Mass.: MIT Press, 1996), 58.

46. De Landa, 28.

47. Ibid., 32.

48. Michel Foucault, "What Is An Author?," *Language, Counter-Memory, Practice: Selected Essays and Interviews*. Ed. Donald F. Bouchard (Ithaca: Cornell University Press, 1977), 124.

49. Among the many textual sites one might locate in order to suggest the substantial overlap that still remains between Deleuzian and Freudian rubrics, it is worth recalling that Deleuze and Guattari anchor the material component of the horizontal axis of assemblage theory in a physiological grounding which trumps vulgar Marxist focus upon production as an explanatory foundation in favor of a biologism amenable to Freudian developmental logics: "We think the

material or machinic aspect of an assemblages relates not to the production of goods but rather to a precise state of intermingling of bodies in a society, including all the attractions and repulsions, sympathies and antipathies, alterations, amalgamations, penetrations and expansions that affect bodies of all kinds in their relations to each other. What regulates the obligatory, necessary, or permitted interminglings of bodies is above all an alimentary regime and a sexual regime" (Deleuze and Guattari, *A Thousand Plateaus*, 90). One might translate such a declaration of the foundational status of alimentary and sexual regimes with Giordano Bruno's declaration that: "The most important of all bonds is the bond of Venus and of love in general, and that which is primarily and most powerfully the opposite of love's unity and happiness is the bond of hate. Indeed, to the degree that we love one of two opposites and contraries of any type, then to that degree we hate and reject the other. These two feelings, or rather, in the last analysis, this one feeling of love (whose substance includes hate) dominates all things, is lord over all things, and elevates, arranges, rules and moderates all things" (Bruno, 174). For a divergent range of perceptive excavations of this border territory as a contemporary methodological problem, see *Deleuze and Psychoanalysis: Philosophical Essays on Deleuze's Debate with Psychoanalysis*, ed. Leen de Bolle (Leuven: Leuven University Press, 2010), passim.

50. Walter Benjamin, *The Origin of German Tragic Drama*, Trans. John Osborne (London: Verso, 1998), 175.

51. Robert Burton, "Definition of Melancholy, Name, Difference" *The Anatomy of Melancholy, What It Is, With All the Kinds, Causes, Symptomes, Prognostics, and Several Cures of It.* Vol. I. Eds. Thomas Faulkner, Nicolas Kiessling, and Rhonda L. Blair (Oxford: Clarendon Press, 1989), Member 3, Subsection I, 162; italics mine.

1. From Dürer's Angel to Harlow's Monkey

1. See Deleuze and Guattari, "1440: The Smooth and the Striated," *A Thousand Plateaus*, 474–500. The opposition between "affect" and "emotion" is omnipresent in the recent flourishing of affect studies, and each particular theorist which one might identify within this group (I have in mind Massumi, Sedgwick, Terada, Ngai, Brennan) produce their own particular parsing of how to articulate the relational disjunction between these two terms. I have opted for the utility of "smooth" and "striated" because, as Deleuze and Guattari deploy them, these terms convey both a difference (they are not of the same nature) and a processual capacity for translation, metamorphosis, and overlap.

2. Leon Battista Alberti, *Della pittura* (On painting), Trans. John R. Spencer (New Haven: Yale University Press, 1956), 77.

3. Ludwig Wittgenstein, *Philosophical Investigations*, Trans. Elizabeth Anscombe (Oxford: Basil Blackwell, 1963), 193–197.

4. Wittgenstein, *Philosophical Investigations*, 209.

5. John Dryden, "Parallel Betwixt Painting and Poetry," *'Of Dramatic Poesy' and Other Critical Essays*, ed. George Watson (London: Dent, 1962), ii, 189.

6. Paolo Giovanni Lomazzo, *A Tracte Containing the Artes of Curious Painting* (1598) (Theatrvm Orbis Terrarvm Ltd./New York: Da Capo Press, 1969), 25.

7. For more on the concept of "the creaturely" as a mode of embodiment that sutures humanity to animal creation, see Eric Santner's *On Creaturely Life: Rilke, Benjamin, Sebald* (Chicago: University of Chicago Press, 2006). For more on the "creaturely" in early modern literature, see Julia Lupton's "Creature Caliban" *Shakespeare Quarterly* 51, no. 1 (2000): 1–24.

8. Benjamin, Walter. *The Origin of German Tragic Drama.* Trans. John Osborne (London: Verso, 1998), 146.

9. As Wilson puts it in his announcement of the term, "A new word is required that conjoins 'black' and 'ecology': melancology, a word in which can be heard the melancholy affect appropriate to the conjunction. [. . .] the plane of immanence of melancology is extinction and nonbeing. All things are destined for extinction; immanent to all being is the irreducible fact of its total negation without reserve or remainder." Yet this landscape also implies the "conceptual personae" of the melancholic who surveys or bemoans its desolation. Scott Wilson, "Introduction to Melancology," *Melancology: Black Metal Theory Symposium II*, 2011.

10. Erwin Panofsky, *The Life and Art of Albrecht Dürer* (Princeton: Princeton University Press, 1943), 170.

11. Klibansky, Saxl, and Panofsky, *Saturn and Melancholy: Studies in the History of Natural Philosophy, Science and Art* (London: Basic Books, 1964), plates 122, 115, 123, and 134. For more on Rosa, see Richard Wallace "Salvator Rosa's *Democritus* and *L'Umana Fragilita*," *The Art Bulletin* 50, no. 1 (March 1968): 21–32.

12. For an account of both the citationality and originality of this image, see Frederick Cummings, "Boothby, Rousseau, and the Romantic Malady." *The Burlington Magazine* 110, no. 789 (December 1968): 659–667. Cumming's provocative argument that the melancholy posture derives from Giulio Campagnola's engraving of *Saturn* would place it as a potential precursor to Dürer, though the difficulty in dating Campagnola's work accurately makes this tentative, and the noted influence of Dürer's work upon at least some of Campagnola potentially complicates this claim.

13. See, for example, the inclusion of Rodin's "Le Penseur" in the catalogue to Jean Clair's *Melancolie: Genie et Follie en Occident* (Paris: Editions Gallimard, 2005), 466.

14. Klibansky, Panofsky, and Saxl, *Saturn and Melancholy: Studies in the History of Natural Philosophy, Science and Art*, passim.

15. Ibid., 320.

16. Plate 61, Paris, 1489 in Klibansky, Panofksy, and Saxl depicts "Melencolye and Entendement" with the propping posture, while the undated medieval

German illuminated manuscript of "The Four Temperaments," Plate 84, assigns the propping posture to "Fleumaticus." *Saturn and Melancholy*, passim.

17. Jane Campbell Hutchison, *Albrecht Dürer: A Biography* (Princeton: Princeton University Press, 1990), 41.

18. See Alfred Winterstein, "Die Quellen zu Dürers 'Melencolia I', *Durer's 'Melancholie' im Lichte der Psychoanalyse* (Leipzig: Internationaler Psychoanalytischer Verlag, 1929), 16–23.

19. A historical and symptomatic disjunction I pursue explicitly in this book in the reading of *The Merchant of Venice*.

20. James Elkins, "Psychomachia," *Pictures of the Body: Pain and Metamorphosis* (Stanford: Stanford University Press, 1999), 71.

21. Wendorf, *Elements of Life*, 89–90.

22. Mary Edmond, *Hilliard and Oliver: The Lives and Works of Two Great Miniaturists* (London: R. Hale, 1983), 85.

23. Roy Strong, *The English Renaissance Miniature* (London: Thames and Hudson, 1983), 142–185.

24. Marcia Pointon, *Hanging the Head: Portraiture and Social Formation in Eighteenth-Century England* (New Haven: Yale University Press, 1993), 4.

25. Ibid., 1.

26. Wendorf, *Elements of Life*, 67.

27. Pointon, 1.

28. Strong, 184.

29. Ibid., 100.

30. Edward Herbert, *The Life of Edward, First Lord Herbert of Cherbury Written by Himself.* J. M. Shuttleworth, ed. (London: Oxford University Press, 1976), 60.

31. Herbert, 44.

32. Ibid., 55.

33. The iconic importance of *The Young Man among Roses* is only exceeded by the amount of debate occasioned by the attribution of the titular young man who was its sitter. Though I share Sir Roy Strong's regard for this work ("an hypnotic image seemingly bearing within itself the quintessence of an age") (Strong, 10), I remain agnostic about his belief that it depicts Essex; what is interesting to me about his argument is that it hinges upon the symbolic interpretation of the image's iconographic content (the man wears black and white and gestures toward eglantine, suggesting a courtly lover's show of allegiance to Elizabeth). That his argument is based upon this kind of content reinforces the extent to which Hilliard, for all his realism in the depiction of jewels and precious metals, is overwhelmingly committed to emblematic modes of representation.

34. Strong, 10.

35. Originally cited in Charles Rosen, "The Ruins of Walter Benjamin," in Gary Smith (ed.), *On Walter Benjamin: Critical Essays and Recollections* (Cambridge, Mass.: MIT Press, 1988), 156.

36. Benjamin, *Origin*, 34.

37. Louis Althusser, *Reading Capital*, trans. Ben Brewster (London: New Left Books, 1970), 187–189.

38. Wendorf, *Elements of Life*, 69.

39. *Remaines*, as quoted by Strong, 96.

40. Benjamin, *Origin*, 87.

41. G. W. F. Hegel, *On the Philosophy of Fine Art*. Trans. F. P. B. Osmaston. Vol. 2 (Oxford: Oxford University Press, 1920), 335.

42. Ibid., 327.

43. The distorted circuit established between body, shield, and verbal crux within this painting confirms James Elkins' observation that "the history and development of heraldic rules and nomenclature might be read as a history of the changing concepts of the human form—medieval heraldry is more organic, and less precisely articulated; Renaissance heraldry is more theoretically elaborated, and prone to distortion; and modern heraldry has largely disintegrated into its component parts (crowns, spheres, banners, and so forth), which have become the simplified logos of companies." Elkins, "Analogic Seeing," *Pictures of the Body: Pain and Metamorphosis*, 205.

44. Thus there is both a resemblance and a crucial difference between the specifically melancholic form of solicitation that I am interested in, and the presentation of the self as to-be-looked-at for which Michael Fried has coined the critical term-of-art "theatricality." "Theatricality," in contrast to "absorption," acknowledges the fact of the viewer directly, often with a direct return of the gaze; melancholic solicitation carries with it an open possibility of skeptical critique about the presence or absence of the humoral condition it models. As a notably "stagey" form of seeming-absorption in an antisocial emotion, melancholy would seem to conceptually short-circuit Fried's opposition completely, but in my reading particular portraits fall upon either side of his conceptual distinction. Some melancholics show up as absorbed (e.g., Rosa's Democritus), while others (surely Hilliard's Herbert) seem very much contained within the theatrical. Michael Fried, *Absorption and Theatricality: Painting and Beholder in the Age of Diderot* (Berkeley: University of California Press, 1980).

45. Ficino, "How Many Things Cause Learned People Either to Be Melancholy or to Eventually Become So" *Three Books on Life*, ed. Clark and Kaske (Tempe: Renaissance Society of America, 2002), 113.

46. Babb, *The Elizabethan Malady*, 74.

47. Harry Berger Jr., *Fictions of the Pose: Rembrandt Against the Italian Renaissance* (Stanford: Stanford University Press, 2000), 173.

48. Ibid., 220.

49. There is a bewildering plenitude of critical texts that develop this notion of early modernity as the preeminent historical site of an "emergent" or proto-contemporary articulation of subjectivity—far too many for me to cite

them all in this space. A smaller number of texts have been either interestingly skeptical of this claim or particularly resourceful in their vindication of it; I have in mind the following: Ronald Levao's *Renaissance Minds and Their Fictions* (Berkeley: University of California Press, 1985); Elizabeth Hanson's *Discovering the Subject in Renaissance England* (Cambridge: Cambridge University Press, 1998); Katharine Eisaman Maus' *Inwardness and Theater in the English Renaissance* (Chicago: University of Chicago Press, 1995); and Christopher Pye's *The Vanishing: Shakespeare, The Subject, and Early Modern Culture* (Durham: Duke University Press, 2000). Unsurprisingly, a concept as richly multivalent as "inwardness" generates multiple explanatory frameworks. Recently, more robustly physiological accounts of how "interiority" might be understood within the period have emerged; notably, David Hillman's *Shakespeare's Entrails: Belief, Scepticism, and the Interior of the Body* (New York: Palgrave Macmillan, 2007) and William W. E. Slight's *The Heart in the Age of Shakespeare* (Cambridge: Cambridge University Press, 2008).

50. Gail Kern Paster, "Melancholy Cats, Lugged Bears, and other Passionate Animals: Reading Shakespeare's Psychological Materialism across the Species Barrier," *Humoring the Body: Emotions and the Shakespearean Stage* (Chicago: University of Chicago Press, 2004), 135–189.

51. Harry F. Harlow, "Self-aggression," *Learning to Love* (New York: Jason Aronson, 1974), 114.

52. Lest this "other than . . ." be taken to concede precisely the issue at stake, I must insist that melancholy abides in a problem space of conflict between its theoretical articulation in Greek medicine (in which the matter of black bile is precisely and utterly "essential") and its expressive multiplicity across bodies and signs. As articulated in the introductory chapter, that problem space is defined for me by the epistemological voids of gaps opened up within melancholy assemblages, a-substantial gaps that exceed or refuse to verify an objective melancholy essence.

53. See Jean Laplanche, "The Kent Seminar," *Seduction, Translation and the Drives*. Trans. Martin Stanton. London: Psychoanalytic Forum, Institute of Contemporary Arts, 1992. I am indebted to Homay King for directing me to this text.

54. For a metaphysical account of the implications of this passage toward death that weds Freud's *Beyond the Pleasure Principle* to a broadly triumphant scientific materialism, see Ray Brassier, "The Truth of Extinction," *Nihil Unbound: Enlightenment and Extinction* (London: Palgrave Macmillan, 2007), 205–240.

55. Jean Clair, *Mélancolie: Génie et Folie En Occident* (Paris: Gallimard, 2005).

56. Christopher Müller, *Bas Jan Ader* (Cologne: Verlag der Buchhandlung Walther Konig, Kunstverein Braunschweig, Bonner and Kunstverein Munchen, 2000), 53–74.

57. Jan Verwoert, *Bas Jan Ader: In Search of the Miraculous* (London: Afterall Books, 2006), 20.

58. See Rene Daalder's documentary "*Here Is Always Somewhere Else: The Disappearance of Bas Jan Ader.*" Rene Daalder, "*Here is always somewhere else: The Disappearance of Bas Jan Ader,*" Cult Epics, 2008, DVD.

2. *Three Hundred Years Out of Fashion*

1. Londré, *Love's Labour's Lost: Critical Essays* (New York: Garland Press, 1997), 45.

2. Ibid., 61.

3. Harold Bloom, *Shakespeare: The Invention of the Human*, 49.

4. Costard's deployment of "remuneration" in conversation with Biron betrays his own desire to take part in courtly language, but it also functions within the play as an economic reading of wit as a kind of inflation of language beyond the gold-standard of referential utility. Instead of the one-to-one correspondence of "three farthings" for three farthings, wit allows Armado to inflate a miserly three farthings into the grandiloquent "remuneration," an act of semantic transformation repeated when Biron transforms a shilling into a "guerdon." As C. L. Barber puts it, "in wit, it is language that gives us something for nothing" (Londré, 155). But Costard's preference for a guerdon to remuneration demonstrates that economic value is not nearly so pliable, and Armado's predilection for piling on phrases bears an inverse correspondence to his actual economic circumstances. Three farthings is, as a tip, rather poor, and, when it is revealed at the close of the play that Armado wears no undershirt, his comparative poverty emerges, placing the compensatory function of wit in darker relief.

5. Hibbard, 110.

6. Traditionally this relationship has been read as a simple matter of contrast and scale. Schlegel's remarks commence what is still today largely the received sense of how to place the sub-plot characters in relation to the court: "The grotesque figures of Don Armado, a pompous fantastic Spaniard, a couple of pedants, and a clown, who between the whiles contribute to the entertainment, are [. . .] well adapted as foils for the wit of so vivacious a society" (Schlegel, *Lectures on Dramatic Art and Literature*, ed. Morrison [London: George Bell and Son, 1879], lecture 24). Like Don Armado himself, the term "foil" says both too little and too much. In assigning "foil" status, Schlegel defines these characters in terms of pure negation; the more grotesque, pompous, or foolish these characters are, the more genteel, articulate, and witty the courtly characters will seem by contrast. But the progress of the play troubles and eventually dissolves such a simple inverse relationship; if, as the play develops, the wit of the court and the pretension and folly of the "grotesques" come to resemble each other,

during the masque of the Nine Worthies, the human dignity of these supposed grotesques as they suffer the court's mockery solicits our sympathy with such force that the contrast is ultimately reversed and transvalued.

7. For more on the complex intersection between humors and class distinctions, see Gail Kern Paster's *Humoring the Body: Emotions and the Shakespearean Stage* (Chicago: University of Chicago Press), 2004.

8. Moncure-Sime, *Shakespeare: His Music and Song*, 48.

9. As quoted in Hibbard, 5.

10. Benjamin, *Arcades*, 64.

11. A closer consideration of how this dialectical theater of fashion operates in *Love's Labour's Lost* could potentially resolve some of the more farfetched critical quarrels made about the proper dating of the play. Consider the following argument made by the Oxfordian Felicia Londré:

> The affinities of *Love's Labour's Lost* with Euphuism [. . .] make 1578 a likely date for the first draft [.] Conversely, the orthodox dating of the play to the 1590s (to make it fit the dates of Shakespeare of Stratford-upon-Avon) posits that the author based his in-jokes upon topics that had been fashionable twelve to fifteen years earlier. How would a young man fresh from a small town have dared to write one of his first plays for and about court society? In fact, how could one who spoke Warwickshire dialect have acquired the verbal facility and sophistication to lampoon a linguistic fad that had flared briefly among courtiers when he was only fourteen? (Londré, 8)

Among the many assumptions made here, the most striking for my purposes is the reliance upon a theory of the "fashionable" as a sort of container from which things fall out (but never back in) at a furious tempo. Even bracketing the assumption that *Love's Labour's Lost* is "about" Euphuism directly and exclusively, this analysis would benefit from a more nuanced account of how social types circulate and are assimilated into a culture, and grow more accessible to satiric examination in popular art precisely as they gain familiarity outside of coteries.

12. Barthes, *The Fashion System*, trans. Matthew Ward and Richard Howard (Berkeley: University of California Press, 1990), ix.

13. Freud, *MM*, 157.

14. Freud's account of the insistent communication of the melancholic is far from typical when compared with other clinical accounts. Working from a definition of melancholia as "unfulfilled mourning for the maternal object," in "The Life and Death of Speech" Julia Kristeva describes the effects of this un-fulfillment on the speech of depressed and melancholic patients:

> Let us keep in mind the speech of the depressed—repetitive and monotonous. Faced with the impossibility of concatenating, they utter sentences that are interrupted, exhausted, come to a standstill. Even phrases they cannot

> formulate. A repetitive rhythm, a monotonous melody emerges and dominates the broken logical sequences, changing them into recurring, obsessive litanies. Finally, when that frugal musicality becomes exhausted in its turn, or simply does not succeed in becoming established on account of the pressure of silence, the melancholy person appears to stop cognizing as well as uttering, sinking into the blankness of asymbolia or the excess of an unorderable cognitive chaos (Kristeva, *Black Sun*, 33).

Kristeva's account of depressive speech seems exactly wrong as a model for the language used by Shakespearean melancholics. Nothing could be further from the language of a Biron or a Hamlet, in which lightning flashes of satiric wit and punning courtly banter alternate with formally gorgeous set pieces. It is hardly surprising that dramatic depictions of love melancholy from the late sixteenth century do not correspond with a contemporary clinical understanding of depression. Shakespeare chose not to sacrifice the flow of dramatic action, not to mention the display of his talent for literary invention, for the sake of clinical accuracy by peopling his stage with alternately silent and tediously monotonous figures. But this disjunction registers more than a merely formal or generic imperative at work. Kristeva's description can help to remind us of the strangeness involved in recognizing Biron or Hamlet as melancholic at all.

15. For more on the notion of "love speech" as a kind of aggressive protestation of feeling which forces effects in and upon the world that resemble but recalibrate the discourse of harm surrounding hate speech, see Owen Ware, "Love Speech," *Critical Inquiry*, Spring 2008, Volume 34. Number Three. University of Chicago Press, 491–509.

16. See William W. E. Slights, "Shakespeare and the Cardiocentric Self," *The Heart in the Age of Shakespeare* (Cambridge: Cambridge University Press, 2008), 149–178.

17. Ferrand, *A Treatise on Lovesickness*, trans. Donald A. Beecher and Massimo Ciavolella (Syracuse: Syracuse University Press, 1990), 256.

18. Judith Butler, *Bodies That Matter: On the Discursive Limits of "Sex"* (New York: Routledge, 1993), 12–13.

19. Slavoj Žižek, *For They Know Not What They Do*, 168.

20. Carla Mazzio, "Acting in the Passive Voice: *Love's Labour's Lost* and the Melancholy of Print," *The Inarticulate Renaissance: Language Trouble in an Age of Eloquence* (Philadelphia: University of Pennsylvania Press, 2009), 142–174.

21. Anne Cheng, *The Melancholy of Race: Psychoanalysis, Assimilation and Hidden Grief* (Oxford: Oxford University Press, 2000), 58.

22. Cheng, 58.

23. Anticipating moves such as Cheng's, Butler's later position has stressed that her theory of performativity was meant to delimit conditions which, as it were, precede the scenario of any given performance:

> . . . performativity cannot be understood outside of a process of iterability, a regularized and constrained repetition of norms. And this repetition is not performed by a subject; this repetition is what enables a subject and constitutes the temporal condition for a subject. This iterability implies that "performance" is not a singular "act" or event, but a ritualized production, a ritual reiterated under and through constraint, under and through the force of prohibition and taboo, with the threat of ostracism and even death controlling and compelling the shape of the production, but not, I will insist, determining it fully in advance (Butler, 95).

Butler seems to want to have it both ways. Constraint is so powerful that it wields the threat of death should resistance be attempted, and constraint already obtains prior to the moment of repetition which it strives to enforce—and yet all is not determined. Eager to quash redemptive scenarios which suggest that freedom from constraint is just a performance away but also trying to suggest that some narrow breathing space of non-determination obtains within the constitutive boundaries of performativity, Butler here draws rhetorical power from the invocation of narratives of punishment (ostracism, death) but at a cost. It would seem that such narratives must necessarily amount to particular "acts" and "events" at some point, and cannot hover menacingly at the level of "temporal conditions" forever if they are to function as effective threats.

24. Pater, *Appreciations: With An Essay on Style* (London: Macmillan, 1868), 167.

25. That said, Moth is not entirely contradicted by this sketch of erotomania; Katherine's enumeration of her sister's symptoms ("melancholy, sad and heavy") serves to confirm, in a darker register, Moth's alleged "parting of sadness and melancholy."

26. Galen, *On the Affected Parts*, trans. Rudolph E. Siegel (Basel: S. Karger, 1976), 131–201.

27. Jackson, *Melancholia & Depression*, 78–114.

28. My use of this figure is of course indebted to the work of Nicolas Abraham and Maria Torok, and in particular their book *The Wolf Man's Magic Word.*

29. This suspension of enjoyment at the mention of the father's name is not entire; the Princess' response "Dead, for my life" also conveys a faint note of satisfaction which articulates a half-conscious thought; it is as if she has blurted out "he is dead so that I may have my life." I am indebted to Donald Friedman for alerting me to this line's wordplay.

30. Abraham and Torok, *The Shell and the Kernel*, Vol. I. Trans. Nicholas T. Rand (Chicago: University of Chicago Press, 1994), 87.

3. Let Me Have Judgment, and the Jew His Will

1. Unless otherwise noted, all quotations from the play are from William Shakespeare, "The Merchant of Venice" *The Complete Works of Shakespeare.* David Bevington, ed. Fifth edition (New York: Longman, 2003).

2. Robert Burton, "Definition of Melancholy, Name, Difference" *The Anatomy of Melancholy, what it is, with all the kinds, causes, symptomes, prognosticks & severall cures of it.* Vol. I. Eds. Thomas Faulkner, Nicolas Kiessling, and Rhonda L. Blair (Oxford: Clarendon Press, 1989), Member 3, Subsection I, 162; italics mine.

3. Jacques Derrida, "Me—Psychoanalysis" *Psyche: Inventions of the Other, Volume I* (Stanford: Stanford University Press, 2007), 130.

4. I invoke the term *case* with due caution. Lauren Berlant's apt definition of the case as "any irritating obstacle to clarity" indirectly recalls Antonio's own self-regard; for my purposes, Berlant goes one better in flagging the response that such irritations call forth ("What matters is the idiom of judgment"), for the means through which Antonio's sadness generates judgment constitutes the therapeutic effect of melancholy for his spectators and would-be caregivers within and without the play (Lauren Berlant, "On the Case," *Critical Inquiry* 33, no. 4, University of Chicago Press, 663).

5. Henry S. Turner, "The Problem of the More-Than-One: Friendship, Calculation, and Political Association in *The Merchant of Venice*," *SQ* 57 (2006): 418. Turner is not by any means to be understood as "hostile" to psychoanalysis (itself a question-begging phrase), but his local flagging of the risks of a reductive form of psychoanalysis in this passage strikes me as symptomatic of a broader problem in the encounter between Renaissance literature and psychoanalytic reading. Of course, the locus classicus for such reactions is Stephen Greenblatt's shot heard round the world, "Psychoanalysis and Renaissance Culture" in *Learning to Curse: Essays in Early Modern Culture* (New York: Routledge, 1990), 131–145. For an account of its destructive wake and a response, see Elizabeth Jane Bellamy, "Desires and Disavowals: Speculations on the Aftermath of Stephen Greenblatt's 'Psychoanalysis and Renaissance Culture,'" *CLIO*, 34, no. 3 (2005): 297–315. Having indicated this chronology, I want to point out that Turner's weary/wary familiarity is not the same as Greenblatt's dismissive rejection: it represents a historical bend in the road in which psychoanalysis is not so much conceptually wrong and therefore worthy of spirited opposition as it is an all too satisfying and overfamiliar resting place for critical practice, a kind of intellectual comfort food which one must forego for the sake of rigor. This is a worse fate than hostility.

6. To be sure, psychoanalytic modes of argument surface in Adelman's text, as when she speaks of Lancelot's "fantasy" of the Christian knave begetting

Jessica as a "condensation"; yet overall, such language is the exception rather than the rule. When considered as a whole, Adelman's readings in *Blood Relations* rely far less upon psychoanalytic concepts than her previous work, and the thrust of the book's arguments are historical and exegetical rather than psychoanalytic. Janet Adelman, "Her Father's Blood: Conversion, Race, and Nation," *Blood Relations: Christian and Jew in The Merchant of Venice* (Chicago: University of Chicago Press, 2008), 72.

7. Disentangling the knot that joins melancholy and masochism cannot proceed unless the tension in application and sense between these two terms always remains in play; their very separateness both within the experience of the play (Antonio begins as a melancholic and ends as a masochist) and within psychoanalytic theory (melancholy is theorized in terms of identification and masochism in relation to libido and the drives) must be drawn out in order for the structure and effects of their productive collision in Antonio to emerge more clearly. Antonio's masochism is not a symptomatic expression of the "previous condition" of his melancholy but rather a revision of it that both repeats and expands its problematic: melancholy's epistemological deadlock of barred but possible meanings gives way to masochism's overproduction of coextensive erotic, juridical, economic, and religious meanings. The first critical reference to Antonio as a masochist that I have discovered occurs as a tantalizing-albeit momentary-reference to Antonio's "intimation of masochistic satisfaction" in René Girard's "To Entrap the Wisest: A Reading of The Merchant of Venice," in *Literature and Society*, ed. Edward W. Said (Baltimore: Johns Hopkins University Press, 1980) 100–119, see esp. 115. Having flagged this possibility, Girard's text moves on to its own concerns with scapegoating and mimesis.

8. See Aristotle "Book XXX: Problems Connected with Prudence, Intelligence and Wisdom," *Problemata.* E. S. Forster, ed. (Vol. VII) (Oxford: Clarendon Press. 1927), 953–955; also Lawrence Babb *The Elizabethan Malady: A Study of Melancholia in English Literature from 1580 to 1642* (East Lansing: Michigan State University Press, 1951), 58–72, see esp. 59. Graziano's remark flags the persistence of the Aristotelian, "genial" definition as an index of unique personal distinction. It seems to me that only such an account can explain the forceful pressure of Graziano's account of melancholy as a bid for status flags the persistence of the Aristotelian, "genial" definition as an index of unique personal distinction (and it is precisely this sort of claim that a purely Galenic understanding of melancholy as merely a chemical imbalance would render unintelligible).

9. As Adelman describes Antonio's "ambivalent desire for this opening up," this speech "images Antonio's body as a container of riches-its own variant of infinite riches in a little room-made visible only by the touch that would annihilate him." Adelman, *Blood Relations*, 118.

10. For more on contemporary debates and the extent and nature of the supposed connection between melancholy and oracular knowledge, see Winfried Schleiner's *Melancholy, Genius and Utopia in the Renaissance* (Wiesbaden: Harrasowitz, 1991). There is also a long sequence of critical writings on the imagination that are potentially relevant here; see Murray Bundy's *The Theory of Imagination in Classical and Medieval Thought* (Urbana: The University of Illinois Press, 1927).

11. Among the many examples, the most direct are: Hamlet's warning to Gertrude: "It will but skin and film the ulcerous place, / Whiles rank corruption, mining all within, / Infects unseen." (Hamlet, 3.4.147–149); Hamlet's image of "th'imposthume of much wealth and peace / That inward breaks, and shows no cause without / Why the man dies" (Hamlet, 4.4.27–29); Polonius' image of "The harlot's cheek, beautied with plast'ring art" (Hamlet, 3.1.50); and Polonius' self-description as "the owner of a foul disease, / To keep it from divulging, let it feed, / Even on the pith of life" (Hamlet, 4.2.21).

12. As William James put it, "the sovereign source of melancholy is repletion. Need and struggle are what inspire and excite us; our hour of triumph brings the void." William James, "Is Life Worth Living?," *The Will to Believe and Other Essays in Popular Philosophy* (New York: Dover Publications, 1956), 47.

13. For more on the "inwardness topos," see Katherine Eisaman Maus, *Inwardness and Theater in the English Renaissance* (Chicago: University of Chicago Press, 1995), 1–34, see esp. 15.

14. See Michael C. Schoenfeldt, "Bodies of Rule: Embodiment and Interiority in Early Modern England," Bodies and Selves in Early Modern England: Physiology and Inwardness in Spenser, Shakespeare, Herbert and Milton (Cambridge: Cambridge University Press, 2000), 1–40, see esp. 21–24.

15. See Cynthia Lewis, "Antonio and Alienation in "The Merchant of Venice," *South Atlantic Review* 48.4 (1983): 19–31; Steve Patterson, "The Bankruptcy of Homoerotic Amity in Shakespeare's *Merchant of Venice*," *SQ* 50 (1999): 9–32; and Lars Engle, " 'Thrift Is Blessing': Exchange and Explanation in *The Merchant of Venice*," *SQ* 37 (1986): 20–37.

16. For an account of moral masochism, see Sigmund Freud, "The Economic Problem of Masochism," in *The Standard Edition of the Complete Psychological Works of Sigmund Freud*, ed. and trans. James Strachey et al., 24 vols. (London: The Hogarth Press, 1953–74), 19:159–170, see esp. 160.

17. Richard Halpern, *Shakespeare Among the Moderns* (Ithaca: Cornell University Press, 1997), 187.

18. Ibid., 197.

19. I sense here a conceptual linkage between Antonio's melancholic pleasure in the truth of castration and what Lacan termed the hysteric's discourse. Alenka Zupancic explains, " 'That is not it!' is the well-known motto of hysterics when it comes to matters of satisfaction, and the other notorious

feature is the emphasis on renunciation, loss, nonsatisfaction, sacrifice. The hysteric is the guardian of the negative, of the incommensurable and the impossible . . . The hysteric is satisfied with nothing, in both possible meanings of this expression." See "When Surplus Enjoyment Meets Surplus Value," in *Jacques Lacan and the Other Side of Psychoanalysis: Reflections on "Seminar XVII,"* ed. Justin Clemens and Russell Grigg (Durham: Duke University Press, 2006), 155–178, see esp. 167.

20. Approaching the nature of "fantasy" from another direction, in his classic essay "Marriage and Mercifixion in *The Merchant of Venice*: the Casket Scene Revisited" Harry Berger, Jr. suggests that perhaps "scenario" is too strong for the productively vague quality of the quasi-plots and virtual outcomes that the play ceaselessly generates, and which he evocatively terms: "the deep structure of latent or tacit action which characterizes the various power struggles. [. . .] It would not be accurate to call them plots or scenarios, because they unfold at a less conscious level than that which we normally associate with the construction of plots and scenarios. This tacit quality is what makes The Merchant of Venice so haunting and tantalizing a play." Harry Berger, "Marriage and Mercifixion in *The Merchant of Venice*: The Casket Scene Revisited" *SQ* 32 (1981): 155–162, 157; italics mine. It is precisely at this "less than conscious" border area that fantasy arises, and I believe that such fantasies are not necessarily tethered to a particular character but to this "deep structure" operative between and among multiple characters. In this limited sense, one could speak of the play, rather than a character, as having a fantasy.

21. J. Laplanche and J-B Pontalis, *The Language of Psychoanalysis*, trans. Donald Nicholson-Smith (New York: W.W. Norton, 1973), 314. Commenting upon this entry, Anne Cheng also notes the scenic or dramatic dimension implicit here: "This is the paradox in Laplanche and Pontalis's formulation: that the fantasmatic signifies a process of desubjectivation, a state of agencylessness, that nonetheless constitutes the subject's sense of integrity and hence his/her potential for agency-a process of scattering the 'self' in order to constitute a stage for the 'self' " Anne Cheng, *The Melancholy of Race: Psychoanalysis, Assimilation, and Hidden Grief* (Oxford: Oxford University Press, 2001), 120; italics mine.

22. Cutting back upon the term's "expansionist tendencies," Torok argues persuasively for a reduced definition of fantasy, in opposition to Susan Isaac's broad definition of fantasy as "the primary content of unconscious mental processes" tout court. The clinical focus of Torok's definition pointedly resists non-therapeutic application; indeed, Isaac's designation of "dramatic representation" as precisely one example of fantasy's forms is particularly the kind of looseness of application that Torok strives to correct. But if Torok shifts the goalposts from a definition of fantasy in terms of its psychical location to a phenomenological account of what it is like to experience the fantasy (of another), in the process she opens up the possibility of the application of her

account of clinical experience to other kinds of experience, dramatic representation included. Nicolas Abraham and Maria Torok, *The Shell and the Kernel*. Vol. i. Nicholas T. Rand, ed. (Chicago: University of Chicago Press, 1994).

23. Ibid., 36.

24. In activating this slippage from "seeing" to "being seen," Shakespeare is building upon a certain grammatical ambiguity already present in the source tale, "Il Pecorone": "Ansaldo besought him to delay his death a few days so that if his Giannetto returned, he could at least see him" (John Russell Brown, ed., *The Merchant of Venice* [London: The Arden Shakespeare, 1955], 148).

25. Richard Krafft-Ebbing, *PsychoPathia Sexualis: A Medico-Forensic Study* (New York: Putnam, 1965), 128; italics mine.

26. Ibid., 153.

27. Gertrude Lenzer, "On Masochism: A Contribution to the History of a Phantasy and Its Theory" *Signs: A Journal of Women in Culture and Society*. Vol. I. (1975) 277–324, see esp. 280.

28. In a suitably comic denouement, the psychoanalytic appropriation of literature concludes in a literary appropriation of psychoanalysis, for if the exemplary case of masochism threatens to dissolve Krafft-Ebbing's science of sexual pathology into a précis of the novels which inspire it, the distorting effects of narrative's gravitational pull led in turn to *Psychopathia Sexualis*'s notorious circulation as a form of erotic literature in its own right, with its case histories functioning as a kind of "Arabian Nights" of perversion.

29. For more on the role of suspense within masochistic fantasy, see Theodor Reik's remarks on the relationship between suspense and end-pleasure in *Masochism in Sex and Society*, trans. Margaret H. Beigel and Gertrud M. Kurth (New York: Grove Press, 1962). See also Victor N. Smirnoff, "The Masochistic Contract," *Essential Papers on Masochism*, ed. Margaret Hanly (New York: New York University Press, 1995), 62–75.

30. Krafft-Ebbing, 182; italics mine. Analogies are not equations; were Krafft-Ebbing's comparison reversible it would render all of Elizabethan London's apprentices, debtors and bondsmen masochistic, which is surely not his intention, or mine. But what is crucial in this assertion of family resemblance between one form of domination and another is not only its promiscuous capacity for slippage from the economic to the sexual and back again, or the mutuality with which it implicates both participants as necessary components within masochistic narrative, but the particularly perverse manipulation of the universality of contractual form, the twist within neutrality, that it makes possible.

31. For an account of the changing role of contract in seventeenth-century culture, see Victoria Kahn's *Wayward Contracts: The Crisis of Political Obligation in England, 1640–1674* (Princeton: Princeton University Press, 2004), see esp. 58–67. Though Krafft-Ebbing and the specifically masochistic contract are notably

absent, Kahn's persistent pressure upon what she terms "the specter of irrational subordination" identifies the lurking question about the motivating passions that drive formalized displays of obedience to the terms of contract; early modern subjects noted that "carried to its logical and perhaps affective extreme, the contractual account of obligation implied the legitimacy of voluntary enslavement for reasons of 'fear, hope, love or any other passion'" Kahn, 58. In *The Merchant of Venice*, Antonio's masochism constitutes one such "other passion" that saturates the bond between creditor and debtor.

32. The pall that hangs over his sentenced conversion to Christianity can in this sense be read as the revenge of the very epistemological deadlock which, at the beginning of the play, hovered around the interpretation of Antonio's melancholy as a blind-spot in social knowledge: at the end of the play, Shylock must not only join Antonio's faith, but, as a converso whose full membership in that faith will always be open to question, Shylock must submit to the same radically unverifiable speculation about what his interiority conceals that Antonio's melancholy already endures.

33. Here we can perhaps venture a fuller understanding of Bassanio's relation to this encounter: he is the occasion for Antonio's entry into the contract, the longed for collaborator whose spectatorial passivity—even impotence—finds its inversion in the violently energetic activity of Shylock. Shylock's cruelty becomes a means by which Antonio can amplify the imagined love and tenderness of Bassanio for his suffering body. But the solicitation of concern is not all. The scenario Antonio is scripting with Shylock seems designed to produce a certain kind of helplessness in Bassanio which ports subjection outward to this viewer, bringing him into affective alignment with the scene he watches. This very helplessness becomes a means of forcing Bassanio to suffer a kind of mimetic desire to take Antonio's place, and thus to avow the same feeling as his friend.

34. Sigmund Freud, "The Economic Problem of Masochism," *Standard Edition*. Trans. James Strachey. Vol. 19. 1961. 160.

35. Ibid., 161.

36. Ibid., 162.

37. Ibid., 161; italics mine.

38. For an analysis of the divergent consequences of this gendered reading of masochism as an imaginary occupation of "the feminine position" for both male and female masochists, see Kaja Silverman's essay "Masochism and Male Subjectivity" *Male Subjectivity at the Margins* (London: Routledge, 1992), 185–214.

39. M. M. Mahood, ed., *The Merchant of Venice* (Cambridge: Cambridge University Press, 1987), 139.

40. Slavoj Žižek, "Is There a Cause of the Subject?," *Supposing the Subject* (London: Verso, 1994), 102.

41. Freud, 165.

42. Ibid., 170.

43. In the context of a discussion of identity theory, Slavoj Žižek has articulated this constitutive entanglement of the rule of law with the threat of violence: "We can sense this concealed dimension of violence already apropos of the everyday "spontaneous" reading of the proposition "law is law"-is not this phrase usually evoked precisely when we are confronted with the "unfair," "incomprehensible" constraint that pertains to the law? In other words, what does this tautology effectively mean if not the cynical wisdom that law remains in its most fundamental dimension a form of radical violence which must be obeyed regardless of our subjective appreciation?" (Slavoj Žižek, *For They Know Not What They Do: Enjoyment as a political factor* [London: Verso, 1991], 34).

44. First and foremost in subtlety and complexity, I have in mind Julia Lupton's consideration of the place of Paul in his own time and in early modernity in her chapter "Citizen Paul" from *Citizen Saints: Shakespeare and Political Theology* (Chicago: University of Chicago Press, 2005), 19–48, and her return to this intersection in "Paul Shakespeare" from *Thinking with Shakespeare: Essays on Politics and Life* (Chicago: University of Chicago Press, 2011), 219–247. In the broader background is the sequence of interpretation leading from Taubes to Badiou and beyond.

45. Shylock's status at the margins of legal protection, and the troubling threat he poses to the integrity of the civil order that both rejects and subjects him, can be usefully compared with Giorgio Agamben's remarks on the place of the refugee in relation to the rhetoric of human rights both guaranteed and delimited by membership in the nation state: "Rights [. . .] are attributed to the human being only to the degree to which he or she is the immediately vanishing presupposition (and, in fact, the presupposition that must never come to light as such) of the citizen" (Giorgio Agamben, "Beyond Human Rights" *Means Without End: Notes on Politics*, trans. Vincenzo Binetti and Cesare Casarino [Minneapolis: University of Minnesota Press, 2000], 20).

46. See Matthew Biberman's reading of Balthasar's "intellectual manipulation of the emergent structures of modernism" in "The Theo-Sexual Matrix," *Masculinity, Anti-Semitism and Early Modern English Literature* (Aldershot: Ashgate, 2004), 39.

47. Gilles Deleuze, "Coldness and Cruelty," *Masochism* (New York: Zone Books, 1991), 82.

48. As Carl Schmitt puts it, "Every concrete juristic decision contains a moment of indifference from the perspective of content, because the juristic decision is not traceable in the last detail to its premises and because the circumstance that requires a decision remains an independently determining moment." Carl Schmitt, "The Problem of Sovereignty as the Problem of the

Legal Form and of the Decision," *Political Theology*, trans. George Schwab (Chicago: University of Chicago Press, 1985), 16–35, see esp. 30.

49. Deleuze, *Masochism*, 88.

4. *That Within Which Passes Show*

1. See Margreta de Grazia, *Hamlet without Hamlet* (Cambridge: Cambridge University Press, 2007), 7–22.

2. These are not, by any means, the only philosophical responses which this oft-cited passage has inspired; see, for instance, Sarah Beckwith's Cavellian reading of the epistemological problem of embodiment in Sarah Beckwith, "The Mind's Retreat from the Face," *Shakespeare and the Grammar of Forgiveness* (Ithaca: Cornell University Press, 2011), 15–35, esp. 16–19.

3. The topology of melancholy as a hidden liquid substance which causes bodily effects but remains encrypted from sight within the solid surface of the body provides a resolutely physiological, material poetic through which Elizabethans metaphorically routed their understandings of the epistemological detective work of intersubjective interpretation, self-knowledge, and knowledge of others. Insofar as that physiological scheme was actively undergoing revision as medicine shifted from a textual discipline to an observational science, it was uniquely pliable as it shifted from literal to metaphorical registers and back again. Melancholy was not the site of a full-blown Cartesian mind/body duality, but rather of a distinctly pre-Cartesian holistic in which bodies and selves were imagined in material terms as subject to the play of humoral forces, and minds were understood as both the effects of those bodies and also as capable of exerting self-disciplinary control and making choices which could impact upon that body for good or ill. There is a "thingliness" to the early modern notion of melancholy as simultaneously a mode of embodiment and a mode of proto-psychological sensibility which is violated if we assimilate Hamlet into the Cartesian project, but there is an implicitly epistemological orientation to the encounter with melancholy which self-displays such as Hamlet's induce. "Melancholy" is a mode of being that is also a mode of knowing, and vice versa. For more on the particulars of this poetic, see Michael C. Schoenfeldt's *Bodies and Selves in Early Modern England: Physiology and Inwardness in Spenser, Shakespeare, Herbert, and Milton* (Cambridge: Cambridge University Press, 1999).

4. Freud, *MM*, 152.

5. Ibid., 153.

6. Ibid.; italics mine.

7. See Ernest Jones' book *Hamlet and Oedipus* (London: Norton, 1949) and Ellie Sharpe's essays "The Impatience of Hamlet" and "An unfinished paper on

Hamlet: Introduction and Extracts," both in *Collected Papers on Psychoanalysis*, ed. Marjorie Brierly (London: Hogarth Press, 1950).

8. Freud, *MM*, 154.

9. Ibid., 155.

10. Ibid.

11. I am indebted here to Janet Adelman's essay "Man and Wife is One Flesh: Hamlet and the Confrontation with the Maternal Body" for my understanding of the complex challenges to identification with the father which Hamlet and Hamlet make legible to us; her reading of the Hyperion image ("Finally, the myth of the father as Hyperion cannot be sustained; and its collapse returns both father and son to the contaminated maternal body," Adelman, 22) informs my own more local claim here. Janet Adelman, "Man and Wife is One Flesh: Hamlet and the Confrontation with the Maternal Body," *Suffocating Mothers: Fantasies of Maternal Origin in Shakespeare's Plays, Hamlet to The Tempest* (New York: Routledge, 1992), 11–38.

12. Eliot, "Hamlet and His Problems," *The Sacred Wood*, 103.

13. Ibid., 102.

14. Ibid., 103.

15. Wilson, *What Happens in Hamlet*, 229.

16. Ibid., 227.

17. The evidence of the Diary certainly suggests the importance of *Hamlet* for Pepys; though Barker does not mention this, on the 13th of November, 1664 Pepys memorized the "to be or not to be" speech, and records having had it set to music (Pepys, *Diary*, 13 November 1664).

18. Barker, *The Tremulous Private Body*, 37.

19. Agamben, *Stanzas*, 3–10; de Grazia, passim.

20. I find I can only understand such an application in a way that is distinctly metaphorical, and translates the economic into the discursive in a rather torturous manner. Insofar as Hamlet's melancholy produces interpretation through the labor of others, he could be said to own a "means of production" of a fanciful kind. But surely such special pleading should not be necessary, and few would find this satisfying as an act of Marxist analysis, (or analysis at all). But Barker, having provocatively lodged Hamlet as the halfway point between the Jacobean and the Pepysian moments in the evolution of a classed and disciplined subject, is the one who must demonstrate in what sense Hamlet incarnates a specifically bourgeois set of values or perspective. Rather like the character he analyses, he gives us only teasing, promissory gestures. One would expect that the exemplary Shakespearean bourgeois subject would be Timon rather than Hamlet.

21. *Hamlet*, I, ii, 64–66, Wofford, ed.

22. *Hamlet*, I, ii, 64–66, Wells and Taylor, ed.

23. Dessen and Thomson, *A Dictionary of Stage Directions in English Drama: 1580–1642*, 15–16.

24. Hunter, 83.

25. Ibid., 84.

26. Merleau-Ponty, *Phenomenology of Perception*, 185.

27. *Hamlet*, I, ii, 64–66, Wells and Taylor, eds.

28. "audience," *OED*.

29. Hirsh, 115–136.

30. For further discussion of this scene and the problem of Gertrude's guilt in relation to the apparition of the ghost, see Drew Daniel, "Neither Simple Allusions Nor True Mirrorings: Seeing Double with Carl Schmitt" *TELOS, Special Issue on Carl Schmitt's Hamlet or Hecuba*, no. 153 (Winter 2010): 51–69.

31. Lemnius, *Touchstone of Complexions*, 21.

32. In enumerating this multitude, the texts differ slightly but to similar overall effect. Q1 gives "Enter King, Queene, Leartes, Lordes," Q2 gives "A table prepard, Trumpets, Drums and officers with Cushions, King, Queen, and all the state, Foiles, daggers, and Laertes," and F1 gives "Enter King, Queene, Laertes and Lords, with other Attendants with Foyles, and Gauntlets, a Table and Flagons of Wine on it" (Kliman, 256–257).

33. This dynamic has been in place since Claudius' own, equally open-ended address to the court that commences the match after the trumpets sound: "You, the judges, bear a wary eye" (5.2.261). This ironic doublespeak is typical of Claudius—of course he does not want the judges to bear too wary an eye, lest they discover the rigged nature of the combat. In effect, Claudius' admonition works to secure and demonstrate the very limits of spectatorial vision by showing it up as reliably unreliable.

34. Lupton and Reinhard, *After Oedipus: Shakespeare in Psychoanalysis*, 17.

35. Lupton and Reinhard's formulation of "Hamlet's legacy" as a "layered identification" (Lupton and Reinhard, 17) articulates not only the, as it were, vertical transmission from Old Hamlet to Hamlet to the reader/spectator, but also the possibility of other lateral slippages in identification between the acknowledged parent Old Hamlet and the hated rival Claudius. In "Man and Wife Is One Flesh," Janet Adelman has discussed the plague of similarity attendant upon Hamlet's attempts to keep Hyperion and the satyr apart, and argues that their separation must ultimately collapse through the contaminating effects of the mother's body (Adelman, 12–13).

36. In a related but distinct argument, John Guillory has argued that the play's digested versions of Montaigne's worldly continental skepticism was designed to pander to the disaffected intellectual circles sympathetic to the cause of Essex. This reading finds a certain limited/tangential support in the facility with which Shakespeare here engages classical example. See John Guillory " 'To Please The Wiser Sort': Violence and Philosophy in *Hamlet*" in *Historicism,*

Psychoanalysis, and Early Modern Culture, ed. Carla Mazzio and Douglas Trevor (New York: Routledge, 2000), 82–110.

37. See Drew Daniel " 'I am more an antique Roman than a Dane': Suicide, Masculinity, and National Identity in *Hamlet*," *Identity, Otherness, and Empire in Shakespeare's Rome*, ed. Maria del Sapio Garbero (Surrey: Ashgate, 2009), 75–91.

5. *Rhapsodies of Rags*

1. I am not the only scholar to have detected critical relationships between Robert Burton and Walter Benjamin; in the introductory chapter to *Affective Mapping: Melancholia and the Politics of Modernism* (2008), Jonathan Flatley also notes the "similarly dialectical mode of argument" shared by these authors. Though his archive concentrates on modernist literature and mine on early modern texts, there is a shared sense of the processual nature of melancholic textuality, and of "melancholizing" as an activity, which unites Flatley's analysis and my own. Jonathan Flatley, *Affective Mapping: Melancholia and the Politics of Modernism* (Cambridge, Mass.: Harvard University Press, 2008), 37.

2. Buck-Morss, *The Dialectics of Seeing*, 50–51.

3. Burton, *AM*, Part III, Section IV, Member I, Subsection I, 332.

4. Benjamin, *Arcades*, 459.

5. Burton, "Democritus to the Reader," 12; italics mine.

6. Burton, *AM*, "Democritus to the Reader," 4.

7. Sontag, *Under the Sign of Saturn*, 111. Both highest and heaviest planet in the ancient cosmology, Saturn registers his power over his children as a crushing gravity, a paternal legacy which Albertus Magnus, in his *Book of Secrets* (1545) glosses thus: "Saturn is the highest planet, whose nature is cold and dry, whose complexion is melancholic, an enemy to mankind, masculine, of the day, evil disposed, and counted the greater misfortune. He is of slow motion, for he performeth his course but in thirty years. [. . .] If he be the Lord of the nativity, he maketh the children of proud heart, lofty in honours, sad, keeping anger, upright in counsel, disagreeing with their wives, malicious; of stature lean, pale, slender, and hard favoured, thick lips, wide nostrils, and cold of nature." Albertus Magnus, *The Books of Secrets of Albertus Magnus, of the Virtues of Herbs, Stones, and Certain Beasts*, ed. Michael R. Best and Frank Brightman (Oxford: Clarendon Press, 1973), 66.

8. Burton, *AM*, "Democritus to the Reader," 7.

9. O'Connell, 35.

10. Benjamin, *Arcades*, 458.

11. Benjamin, *Origin*, 157.

12. Raymond Williams "Structures of Feeling," *Marxism and Literature* (Oxford: Oxford University Press, 1977), 128–136.

13. Fish, *Self-Consuming Artifacts*, 304.

14. Ibid., 349.

15. Fox, *The Tangled Chain*, 13.

16. Lund has the honesty to grant the difficulty in assessing the outcomes of this therapeutic agenda; in her conclusion, having surveyed the marginalia in over twenty-five copies of the *Anatomy*, she acknowledges that "There is no trace in the evidence I have surveyed of readers who used the work to treat their own or others' melancholy, for instance, or who found reading it a therapeutic process. This is not to say, however, that they did not do so." Mary Ann Lund, *Melancholy, Medicine and Religion in Early Modern England: Reading the Anatomy of Melancholy* (Cambridge: Cambridge University Press, 2010), 202. The porosity of the *Anatomy* to unverifiable cycles of personal consultation and reuse is just one local example of broader conditions of textual circulation generally. To contribute to Lund's archive, I will confess that the vigorously underlined passages in my own 1651 copy of the *Anatomy* suggest that its previous reader, one Martha Tunstall, felt moved to record her progress across its pages.

17. Burton, *AM*, "Democritus to the Reader," 6.

18. Burton, *AM*, "Democritus to the Reader," 11. See Mary Thomas Crane's *Framing Authority: Sayings, Self, and Society in Sixteenth Century England* for a sensitive, thorough analysis of both the commonplace book and the broader cultural context of "framing and gathering" in the period.

19. Burton, *AM*, "Democritus to the Reader," 11.

20. See Jonathan Sawday, "Shapeless Eloquence: Burton's Anatomy of Knowledge," *English Renaissance Prose: History, Language and Politics*, ed. Neil Rhodes (Tempe: Arizona State University Press, 1997), 173–202. See also David Renaker, "Robert Burton and Ramist Method," *Renaissance Quarterly* 24, no. 2 (Summer 1971): 210–220.

21. Babb, *Sanity in Bedlam*, 10.

22. Indeed, one need only point out that Sir Thomas Browne's elaboration of the quincunx in *The Garden of Cyrus* (1658), if it does not swell as abundantly as Burton's text, is nonetheless grown from a more minuscule seed of discourse.

23. Flatley, *Affective Mapping: Melancholia and the Politics of Modernism* (Cambridge, Mass.: Harvard University Press, 2008), 37.

24. Adam Kitzes, *The Politics of Melancholy from Spenser to Milton* (London: Routledge, 2006), 150.

25. Burton as quoted in Evans, 55.

26. Burton, *AM*, "Democritus to the Reader," 18.

27. Benjamin, *Origin*, 232.

28. The oddly inorganic intrusion of this subsection is explained in passing by Ruth Fox; it was added by Burton in the third edition (Fox, 7).

29. Burton, *AM*, Part I, Section III, Member III, Subsection I, 418.

30. Ibid., 419.

31. Burton, *AM*, Part II, section II, Member VI, subsection I, 119; Babb, *Sanity in Bedlam*, 44.

32. I am drawing a distinction here between the perspective of melancholy as it is modeled by Burtonian "melancholic scholarship," melancholy as it is attested to in the past (as in Burton's assertion in the preface that in the past he was troubled by melancholy), and this direct attempt to imagine what it is like to feel melancholy.

33. Burton, *AM*, Part I, Section III, Member III, Subsection I, 420.

34. Evans, *The Psychiatry of Robert Burton*, 51.

35. Part of the representational problem occasioned by melancholy is that it is both "thingly" (an object, in this case a liquid mass) and hidden, occulted—because it is an internal substance which could not be seen and touched directly, even in its material form it was always absent, notional, a hypothesis. For more on the conceptual problems in representing the bodily interior, see the chapter "The Uncanny Body" from Jonathan Sawday's *The Body Emblazon'd: Dissection and the Human Body in Renaissance Culture*, 141–182.

36. Burton, *AM*, Part I, Section III, Member III, Subsection I, 421.

37. See Gill Speak, "An Odd Kind of Melancholy: Reflections on the Glass Delusion in Europe (1440–1680)," *History of Psychiatry* 1, no. 2 (June 1990): 191–206.

38. David R. Parker, *The Commonplace Book in Tudor London* (New York: University Press of America, 1998), 44.

39. Burton, *AM*, Partition III, Section IV, Member II, Subsection VI, 443.

40. Burton, *AM*, lxix.

41. Shilleto edition, Burton, *AM*, Vol. I, 485.

42. Burton, *AM*, Part I, Section III, Member III, Subsection I, 421.

43. Ibid.

44. To borrow a phrase from Deleuze, Burton's "line of flight" defines a kind of improvisatory spiral in which he rotates around a fixed point (melancholy) at constantly variable lengths of digression and focus.

45. The particular esteem of these two authors for Burton is casually indexed in their yoking together in his apolegetic "Conclusion to the Reader" which appended the 1621 edition of the *Anatomy*: "no man can observe all, much is defective, and may be justly taxed, altered in Galen, Aristotle, and the very best" (Kiessler, ed., vol. III, 471; italics mine). Here Burton both grants himself license to respond critically and demonstrates the exemplary status of these *auctors*.

46. Benjamin, *Origin*, 150.

47. Lauren Shohet, "YouTube, Use, and the Idea of the Archive" *Shakespeare Studies*, Volume XXXVIII (Madison: Fairleigh Dickinson University Press, 2010), 68–76; see esp. 73.

48. Burton's tendency to Latinize English names (Bergen Evans notes that the homely John Holywood becomes in Burton's text the grandiose and mysterious Sacrobosco) is a way of downplaying vernacular English sources and adding cosmopolitan flair, but it also furthers his anachronistic leveling down of the difference between ancient and modern by lending an ersatz antiquity to a contemporary writer, and thus adds to the anachronistic effect (Evans, 30).

49. Wood, I, 628; italics mine.

50. Faulkner and Kiessling introduction, Burton, *AM*, xxxviii.

51. Burton, "Democritus to the Reader," 20.

52. Babb, *Sanity in Bedlam*, 24.

53. Benjamin, *Origin*, 13.

54. Ibid., 27.

55. *Natur-Geschichte* is a richly nuanced Benjaminian term of art. In it Benjamin tries to convey the dialectical relationship between the natural and historical which can be critically realized through, on the one hand, a kind of pitiless gaze at cultural productions whose supersession and fall into "pastness" have lent them the petrified immobility of a fossil, and a more philosophically ambitious attempt to investigate the structurally timeless law revealed in a moment of "ur-phenomena," a kind of basic ontological moment in which structure and its material expression are united and coextensive. For more on this concept, see Beatrice Hanssen's *Walter Benjamin's Other History: Of Stones, Animals, Human Beings, and Angels*. Berkeley: University of California Press, 2000.

56. Benjamin, *Origin*, 27.

57. Ibid., 28.

58. Ibid.

59. Ibid.

60. Ibid., 29.

61. O'Connell, 88.

62. Ibid.

63. Boswell, I, 389.

64. Deleuze, *Two Regimes of Madness*, 177.

65. Benjamin, *Arcades*, 527.

66. Benjamin as quoted by Agamben, *The Man Without Content*, 104.

67. Benjamin, *Origin*, 28.

68. Ibid., 34.

69. Althusser, 187–189.

6. *My Self, My Sepulcher*

1. Burton, *Anatomy of Melancholy*, I, 162.

2. Participants in the *Samson Agonistes* "debate" (if that is what it is/was) are not arguing coherently from shared terms, but collaboratively dismantling the

platform upon which such a debate might take place by insisting upon mutually incongruous semantic constructions of "terrorism," with Carey's decision to elastically apply a modern political term to a Renaissance poem from a Biblical source cue-ing a nominalist restriction of the phrase on the part of Fish. Since Carey's accusation of terrorist sympathy is phrased as a counterfactual (i.e., "if this reading is right, then Milton was a terrorist, so this reading must be wrong") and Fish's historicist countermove puts terrorism out of rhetorical reach (i.e., "it's not meaningful to use the word terrorism here"), both parties are in fact engaged in the same protective gesture with respect to the political innocence of the Miltonic corpus. See Feisal G. Mohamed's "Confronting Religious Violence: Milton's 'Samson Agonistes.'" *PMLA* 120, no. 2 (2005): 327–340 for a full explication of the buried conceptual consensus that stands behind this alleged debate, and, for a further explication, his *Milton and the Post-Secular Present: Ethics, Politics, Terrorism* (Stanford: Stanford University Press, 2011).

3. Elizabeth Freeman, "Packing History, Count(er)ing Generations" *New Literary History* 31 (2000): 727–744.

4. See Jeffrey J. Cohen, "The Insistent Return of a Corporealized Historiography and Future Making: Elizabeth Freeman's Erotohistoriography," http://www.inthemedievalmiddle.com/2009/01/insistent-return-of-corporealized.html.

5. For a particularly flexible critical account of the emergent queerness of temporality already implicit in the historical framing of early modernity as such, see Carla Freccero "Prolepses: Queer / Early / Modern" *Queer / Early / Modern* (Durham: Duke University Press, 2006), 1–13.

6. See N. K. Sugimura, *"Matter of Glorious Trial": Spiritual and Material Substance in Paradise Lost* (New Haven: Yale University Press, 2009).

7. Descartes, *The Passions of the Soul*, 36.

8. Babb, *The Elizabethan Malady*, 1.

9. Spinoza, *Ethics*, Book III, Scholium to Proposition 11, p. 110.

10. Bailey, *Flaunting*, 48.

11. The "yoke" invoked in the phrase "the yoke of foul effeminacy" is itself a kind of rhetorical switch point. Here it figures not simply the subjection of man to woman, hence a reversal of the expected hierarchical power relations within marriage, but also the yoking of manhood to effeminacy. This "misyoking marriage" is a grotesque inversion of the positive ways in which marital submission is depicted as a just expression of manly self-discipline in the divorce tracts, as when Milton alleges that if Parliament wisely permits divorce to justly dissatisfied husbands, then "the yoke of prudent and manly discipline will be generally submitted to." (Milton, *The Doctrine and Discipline of Divorce*, 863.)

12. "effeminate," *OED*; italics mine.

13. Halperin, *How to Do the History of Homosexuality*, 33.

14. Greenblatt, "Fiction and Friction," *Shakespearean Negotiations*, 73. I do not mean to suggest that this reading of the "One-Sex Model" hegemonically determines the current state of early modern scholarship about gender. The overly rigid application of this model has been challenged on both dramatic grounds and on medico-historical grounds from several directions; see Janet Adelman's essay "Making Defect Perfection: Shakespeare and the One-Sex Model" in *Enacting Gender on the English Renaissance Stage*, edited by Viviana Comensoli and Anne Russell, for an alternative to the Laqueur/Greenblatt formulation.

15. Kahn, "Aesthetics as Critique: Tragedy and Trauerspiel in *Samson Agonistes*," 104–129.

16. Milton, *Eikonoklastes* (1649), 1061.

17. Deborah Shuger, *Habits of Thought in the English Renaissance: Religion, Politics, and the Dominant Culture* (Toronto: University of Toronto Press, 1997), 254.

18. For an account of the rhetoric of ipseity in relation to self-killing, see Eric Langley "Visions of Narcissus," *Narcissism & Suicide in Shakespeare and his Contemporaries* (Oxford: Oxford University Press, 2009), 27–53.

19. Spinoza, *Ethics*, Book III, Prop 55, p. 135.

20. Freud, *MM*, 246.

21. Ibid., 247.

22. Aristotle, *Metaphysics*, Book Kappa, 9, 1066a, 20.

23. Ibid., 1065b, 30.

24. Sugimura, 280.

25. Inwood and Gerson, *The Stoics Reader*, 166.

26. The commentary of Laura Lunger Knoppers for the Oxford University Press edition of *The 1671 Poems* reduplicates this conceptual divergence in the "or" that sunders her diplomatically equivocal gloss on the word "motion": "an inner prompting or impulse; *or* a prompting or impulse originating from God, especially a working of God in the soul." John Milton, *The Complete Works of John Milton, Vol. II: The 1671 Poems: Paradise Regain'd and Samson Agonistes*, ed. Laura Lunger Knoppers (Oxford: Oxford University Press, 2008), 156; italics mine.

27. It would seem that this indeterminacy draws its source from Calvin's own insistence upon the difficulty in distinguishing the elect from the reprobate "from inside," as it were; Calvin admits that "there be a great likeness and affinitie betwene the elect of God, and them that are endued with a fallinge faith for a time" (*Institutes*, 3.2.11).

See Stachniewski's chapter "English Puritanism and the Social Reality of Religious Despair," and in particular his reading of the effect of the *Institutes* upon the English public, from *The Persecutory Imagination* (1–84, passim).

28. "And how could I deny that these hands and this body belong to me, unless perhaps I was to assimilate myself to those insane persons whose minds

are so troubled and clouded by the black vapours of the bile that they constantly assert that they are kings, when they are very poor' that they are wearing gold and purple when they are quite naked' or who imagine that they are pitchers or that they have a body of glass" (Descartes, *First Meditation*, 96).

29. Gordon Teskey, "Samson and the Heap of the Dead," *Delirious Milton: The Fate of the Poet in Modernity* (Cambridge, Mass.: Harvard University Press, 2006), 200.

30. There are, inevitably exceptions. Some recent cases in point concerning *Samson Agonistes* include, as stated, Gordon Teskey's evocation of a "truthfulness that renders everything uncertain" in the poem, and David Loewenstein's account of the way that the two 1671 poems "jostle against each other" without any clear account of their mutual relationship. See Gordon Teskey, "Samson and the Heap of the Dead," *Delirious Milton: The Fate of the Poet in Modernity* (Cambridge, Mass.: Harvard University Press, 2006), 180–203; David Loewenstein, "From Politics to Faith in the Great Poems?" *Visionary Milton: Essays on Prophecy and Violence*, ed. Peter E. Medine, John T. Shawcross, and David V. Urban (Pittsburgh: Duquesne University Press, 2010), 269–287.

31. Fish, *How Milton Works*, 572.

32. In this, Knott enters an ongoing conversation. In the secondary literature there is a long tradition of critical debate among Miltonists about whether or not Samson's death qualifies as martyrdom. Irene Samuel and Lawrence Hyman do not regard his death in such a light, while Krouse, Chambers, and Harris do (Knott, 171). Knott notes that Samson's fate in the closet drama collapses together all of the possible outcomes that in *Christian Doctrine* Milton finds definitive of martyrdom: "the profession of true religion when it leads to death or imprisonment or torture or disgrace" (Knott, 167).

33. Knott, *Discourses of Martyrdom in English Literature, 1563–1694*, 178.

34. Gordon Teskey, *Delirious Milton*, 182.

35. Sartre, *Sketch for a Theory of the Emotions*, 61.

36. Milton, *PL*, XII, 645.

Epilogue: Disassembling Melancholy

1. For its most recent articulation, see Manuel De Landa, *Philosophy and Simulation: The Emergence of Synthetic Reason* (London: Continuum, 2011).

2. Eve Kosofsky Sedgwick, "Introduction: Axiomatic," *The Epistemology of the Closet* (Berkeley: University of California Press, 1990), 23.

3. Daniel Dennett, *Consciousness Explained* (New York: Little, Brown & Company, 1991), 267.

4. As Darian Leader has pointed out, "According to the World Health Organization, by 2010 depression will be the single largest public health problem after heart disease." Darien Leader, *The New Black: Mourning, Melancholia and Depression* (New York: Graywolf Press, 2008), 13.

5. Damasio, *Looking for Spinoza: Joy, Sorrow, and the Feeling Brain* (New York: Mariner Books, 2003), 27.

6. Ibid., 27.

7. Ibid., 30; italics mine.

8. For a defense of the "inherent rhetoricity of emotion," see Daniel Gross, "Defending the Humanities with Charles Darwin's The Expression of the Emotions in Man and Animals (1872)" *Critical Inquiry* 37, no. 1 (Autumn 2010): 34–60; for a critique of the celebrated emotion research of Ekman, see Ruth Leys, "How Did Fear become a Scientific Object and What Kind of Object is It?" *Representations*, no. 110 (Spring 2010): 66–104.

9. For a usefully critical account of the current state of psychological research into the measurement and study of emotion, see Jerome Kagan, *What Is Emotion?: History, Measures, and Meanings* (New Haven: Yale University Press, 2007). For the "extended mind hypothesis" account of the brain in relation to the surrounding world, see Andy Clark, *Supersizing the Mind: Embodiment, Action, and Cognitive Extension* (Oxford: Oxford University Press, 2008). See also Alva Noe, *Out of Our Heads: Why You Are Not Your Brain and Other Lessons from the Biology of Consciousness* (New York: Hill and Wang, 2009).

10. For an example of a capaciously inclusive but "realist" social ontology, see John Searle's account of "institutional facts." John Searle, *Making the Social World* (Oxford: Oxford University Press, 2010).

11. Hobbes pounced early on, declaring "incorporeal substance" to be "contradictory and inconsistent" in the fourth chapter of *Leviathan*, "Of Speech." Thomas Hobbes, *Leviathan*, Ed. Richard Tuck (Cambridge: Cambridge University Press, 1996), 30.

12. Spinoza, *Ethics*, IIP7.

13. For the strongest recent reversal of Kant's critical turn, see Quentin Meillassoux, "Ptolemy's Revenge" *After Finitude: An Essay on the Necessity of Contingency*, Trans. Ray Brassier (London: Continuum, 2008), 112–129.

14. Graham Harman, "Space, Time and Essence: An Object-Oriented Approach," *Towards Speculative Realism* (London: Zero Books, 2010), 149.

15. Indeed, Harman locates just such a *terra incognita* within the unknowable interiority of each object as such. Graham Harman, "On Vicarious Causation" *Collapse Vol. II: Speculative Realism* (March 2007): 187–221.

16. http://www.metal-archives.com/bands/Striborg/16746.

17. For more on depressive black metal and its "belated" relationship to pioneers within the black metal genre, see Dominic Fox, "A Sermon in the Name of Death," *Cold World: The Aesthetics of Dejection and the Politics of Militant Dysphoria* (London: Zero Books, 2009), 43–57. Though Fox's analysis concentrates upon the American solo artist Xasthur, much of his analysis also illuminates Striborg's practice, habits, and goals.

18. http://www.discogs.com/artist/Striborg.

19. There are other forms of heavy metal which are just as strongly in dialogue with melancholy aesthetics, and a different book might be written to exclusively track this fertile dialogue. Slowed to a lethargic tempo that courts absolute stasis, "doom metal" offers a kind of phenomenologically down-tuned version of heavy metal's generally propulsive thrust, and this sense of physical shutdown is also reflected in the misery, scarcity, and anticipatory dread that often make up doom's lyrical content. Offering a case in point, the doom band Esoteric's song "The Noise of Depression" from the aptly titled album "Epistemological Despondency" articulates rather exactly the interpretive/affective dynamic of melancholy as a problem in and for knowledge: Depression shows up as a kind of "noise" that corrodes the signal of communication as such. Esoteric "The Noise of Depression", *Epistemological Despondency*, Aesthetic Death, 1994.

20. Nicola Masciandaro, "Anti-Cosmosis: Black Mahapralaya," *Hideous Gnosis: Black Metal Theory Symposium I*, 2010, 90.

21. Drew Daniel, "Corpsepaint as Necrominstrelsy, or Towards the Re-Occultation of Black Blood," *Melancology: Black Metal Theory and Ecology* 2013, ed. Scott T. Wilson (London: Zero Books).

22. Striborg, "Digging A Ditch To Die In," *Embittered Darkness/Isle De Morts*, Displeased Records/Southern Lord Recordings, 2006.

23. http://blackmetaltheory.blogspot.com/2010/11/melancology-black-metal-theory.html.

24. See Scott Wilson's edited anthology *Melancology: Black Metal Theory and Ecology*; and Timothy Morton, *Ecology Without Nature: Rethinking Environmental Aesthetics* (Cambridge, Mass.: Harvard University Press, 2007).

25. As I completed this manuscript, Julia Reinhard Lupton shared with me her essay on "Softscapes" for *The Return of Theory in Early Modern Studies 2* (London: Palgrave/Macmillan, forthcoming). In a case of parallel evolution or long-distance sympathy, Lupton has also fashioned a reading of von Trier's film which also designates this "magic cave" as an assemblage. I am happy to cite her reading here as one in resonant sympathy with my own formulations and concerns.

Bibliography

Aasand, Hardin, ed. *Stage Directions in Hamlet: New Essays and New Directions.* London: Associated University Presses, 2003.

Abraham, Nicolas, and Maria Torok. *The Shell and the Kernel.* Vol. 1. Ed. Nicholas T. Rand. Chicago: University of Chicago Press, 1994.

———. *The Wolf Man's Magic Word: A Cryptonymy.* Minneapolis: University of Minnesota Press, 1986.

Adelman, Janet. *Blood Relations: Christian and Jew in* The Merchant of Venice. Chicago: University of Chicago Press, 2008.

———. "Making Defect Perfection: Shakespeare and the One-Sex Model." In *Enacting Gender on the English Renaissance Stage,* edited by Viviana Comensoli and Anne Russell, 23–53. Chicago: University of Illinois Press, 1998.

———. *Suffocating Mothers: Fantasies of Maternal Origin in Shakespeare's Plays,* Hamlet *to* The Tempest. London: Routledge, 1992.

Agamben, Giorgio. "Beyond Human Rights." In *Means Without End: Notes on Politics.* Trans. Vincenzo Binetti and Cesare Casarino. Minneapolis: University of Minnesota Press, 2000. 15–29.

———. *The Man without Content.* Trans. Georgia Albert. Stanford: Stanford University Press, 1994.

———. *Stanzas: Word and Phantasm in Western Culture.* Trans. Ronald L. Martinez. Minneapolis: University of Minnesota Press, 1993.

Alberti, Leon Battista. *Della Pittura (On Painting).* Trans. John R. Spencer. New Haven: Yale University Press, 1956.

Albertus Magnus. *The Book of Secrets of Albertus Magnus, of the Virtues of Herbs, Stones and Certain Beasts.* Ed. Michael R. Best and Frank Brightman. Oxford: Clarendon Press, 1973.

Althusser, Louis. *Reading Capital.* Trans. Ben Brewster. London: New Left Books, 1970.

Arikha, Nogha. *Passions and Tempers: A History of the Humours.* New York: Ecco, 2007.

Aristotle. *Metaphysics*. Trans. Richard Hope. Ann Arbor: The University of Michigan Press, 1952.

———. *Problems*. Trans. W. S. Hett. London: William Heinemann, 1936.

———. *The Works of Aristotle*. Ed. J. A. Smith and W. D. Ross. Oxford: Clarendon Press, 1931.

Ashbery, John. *Some Trees*. New York: Corinth Books, 1970.

Avicenna. *The Canon of Medicine*. Trans. O. Cameron Gruner, M.D. London: Luzac & Co., 1930.

Babb, Lawrence. *The Elizabethan Malady: A Study of Melancholia in English Literature from 1580 to 1642*. East Lansing: Michigan State College Press, 1951.

———. *Sanity in Bedlam: A Study of Robert Burton's Anatomy of Melancholy*. New York: Praeger, 1977.

Badiou, Alain. *Deleuze: The Clamor of Being*. Trans. Louise Burchill. New York: University of Minnesota Press, 1999.

———. *Logics of Worlds*. Trans. Alberto Toscano. London: Continuum, 2009.

Bailey, Amanda. *Flaunting: Style and the Subversive Male Body in Renaissance England*. Toronto: University of Toronto Press, 2007.

Barker, Francis. *The Tremulous Private Body: Essays on Subjection*. London: Methuen, 1984.

Barthes, Roland. *The Fashion System*. Trans. Matthew Ward and Richard Howard. Berkeley: University of California Press, 1990.

Beckwith, Sarah. "The Mind's Retreat from the Face." *Shakespeare and the Grammar of Forgiveness*. Ithaca: Cornell University Press, 2011.

Bellamy, Elizabeth Jane. "Desires and Disavowals: Speculations on the Aftermath of Stephen Greenblatt's 'Psychoanalysis and Renaissance Culture.'" *CLIO* 34, no. 3 (2005): 297–315.

Benjamin, Walter. *The Arcades Project*. Trans. Howard Eiland and Kevin McLaughlin. Cambridge, Mass.: Harvard University Press, 1999.

———. *Illuminations*. New York: Schocken Books, 1968.

———. *On the Origin of German Tragic Drama*. London: New Left Books, 1977.

———. *The Origin of German Tragic Drama*. Trans. John Osborne. London: Verso, 1998.

———. *Reflections*. Trans. Edmund Jephcott. New York: Schocken Books, 1986.

Berger Jr., Harry. *Fictions of the Pose: Rembrandt against the Italian Renaissance*. Stanford: Stanford University Press, 2000.

———. "Mercy and Mercifixion in *The Merchant of Venice*: The Casket Scene Revisited." *Shakespeare Quarterly* 32 (1981): 155–162.

Berlant, Lauren. "On the Case." *Critical Inquiry* 33, no. 4 (2007): 663–673.

Bertram, Paul, and Bernice W. Kliman, eds. *The Three-Text Hamlet: Parallel Texts of the First and Second Quartos and First Folio*. By William Shakespeare. New York: AMS Press, 1991.

Biberman, Matthew. *Masculinity, Anti-Semitism and Early Modern Literature*. Aldershot: Ashgate, 2004.

Bloom, Harold. *Shakespeare: The Invention of the Human*. London: Riverhead, 1999.

Borges, Jorge Luis. *Other Inquisitions: 1937–1952*. Trans. Ruth Simms. Austin: University of Texas Press, 1964.

Borris, Kenneth. *Same-Sex Desire in the English Renaissance: A Sourcebook of Texts, 1470–1650*. London: Garland, 2003.

Boswell, James. *The Life of Johnson*. Vol. 1. London: J. M. Dent, 1949.

Brassier, Ray. *Nihil Unbound: Enlightenment and Extinction*. London: Palgrave Macmillan, 2007.

Bray, Alan. *Homosexuality in Renaissance England*. London: Gay Men's Press, 1982.

Brooke, C. F. Tucker, ed. *English Drama: 1580–1642*. Lexington: D. C. Heath and Company, 1933.

Buck-Morss, Susan. *The Dialectics of Seeing: Walter Benjamin and the Arcades Project*. Cambridge, Mass.: The MIT Press, 1991.

Bundy, Murray. *The Theory of Imagination in Classical and Medieval Thought*. Urbana: The University of Illinois Press, 1927.

Burton, Robert. *The Anatomy of Melancholy, What it is, with All the Kinds, Causes, Symptomes, Prognosticks, and Severall Cures of It. In Three Partitions, with Their Severall Sections, Members, and Subsections, Philosophically, Medicinally, Historically, Opened and Cut Up* (1651). Ed. Rev. A. R. Shilleto. London: George Bell & Sons, 1893.

———. *The Anatomy of Melancholy, What it is, with All the Kinds, Causes, Symptomes, Prognosticks, and Severall Cures of It. In Three Partitions, with Their Severall Sections, Members, and Subsections, Philosophically, Medicinally, Historically, Opened and Cut Up* (1651). Ed. Thomas C. Faulkner, Nicholas K. Kiessling, and Rhonda L. Blair. Oxford: Clarendon Press, 1989.

Butler, Judith. *Bodies That Matter: On the Discursive Limits of "Sex."* New York: Routledge, 1993.

Calvin, John. *Institutes of the Christian Religion*. Vol. 1. Trans. Ford Lewis Battles. Philadelphia: The Westminster Press, 1960.

Cassian, John. *Institutes*. Trans. Boniface Ramsey. New York: The Newman Press, 2000.

Chadwick, John, and W. N. Mann, eds. *The Medical Works of Hippocrates*. Oxford: Blackwell Scientific Publications, 1950.

Charney, Maurice. "Asides, Soliloquies, and Offstage Speech in *Hamlet*: Implications for Staging." In *Shakespeare and the Sense of Performance*, edited by Marvin and Ruth Thompson. Newark: University of Delaware Press, 1987.

Cheng, Anne. *The Melancholy of Race: Psychoanalysis, Assimilation, and Hidden Grief*. Oxford: Oxford University Press, 2000.

Clair, Jean. *Melancolie: Genie et Folie en Occident*. Paris: Editions Gallimard, 2005.

Clark, Andy. *Supersizing the Mind: Embodiment, Action, and Cognitive Extension.* Oxford: Oxford University Press, 2010.

Crane, Mary Thomas. *Framing Authority: Sayings, Self, and Society in Sixteenth-Century England.* Princeton: Princeton University Press, 1993.

Cummings, Frederick. "Boothby, Rousseau, and the Romantic Malady." *The Burlington Magazine* 110, no. 789 (December 1968): 659–668.

Damasio, Antonio. *Looking for Spinoza: Joy, Sorrow and the Feeling Brain.* New York: Harcourt, 2003.

Daniel, Drew. "'I am more an antique Roman than a Dane': Suicide, Masculinity, and National Identity in *Hamlet.*" In *Identity, Otherness and Empire in Shakespeare's Rome,* edited by Maria del Sapio Garbero, 75–91. Surrey: Ashgate, 2009.

———. "Neither Simple Allusions Nor True Mirrorings: Seeing Double with Carl Schmitt." *TELOS, Special Issue on Carl Schmitt's Hamlet or Hecuba* 153 (Winter 2010): 51–69.

Dash, Irene. *Wooing, Wedding, and Power: Women in Shakespeare's Plays.* New York: Columbia University Press, 1981.

De Bolle, Leen, ed. *Deleuze and Psychoanalysis: Philosophical Essays on Deleuze's Debate with Psychoanalysis.* Leuven: Leuven University Press, 2010.

de Grazia, Margreta. Hamlet *without Hamlet.* Cambridge: Cambridge University Press, 2007.

DeLanda, Manuel. *A New Philosophy of Society: Assemblage Theory and Social Complexity.* London: Continuum, 2006.

———. *Philosophy and Simulation: The Emergence of Synthetic Reason.* London: Continuum, 2011.

Deleuze, Gilles. *Masochism.* New York: Zone Books, 1991.

———. *Pure Immanence: Essays on a Life.* Trans. Ann Boyman. New York: Zone Books, 2001.

———. *Two Regimes of Madness: Texts and Interviews 1975–1995.* Trans. Ames Hodges and Mike Taormina. New York: Semiotext(e), 2006.

Deleuze, Gilles, and Felix Guattari. *A Thousand Plateaus: Capitalism and Schizophrenia.* Trans. Brian Massumi. Minneapolis: University of Minnesota Press, 1987.

Dennett, Daniel. *Consciousness Explained.* New York: Back Bay Books, 1992.

Derrida, Jacques. *The Politics of Friendship.* London: Verso, 1997.

———. *Psyche: Inventions of the Other, Volume I.* Stanford: Stanford University Press, 2007.

Descartes, René. *The Passions of the Soul.* 1649. Trans. Stephen Voss. Cambridge, Mass.: Hackett Publishing Company. 1989.

———. *The Philosophical Writings of Descartes.* Vol. 2. Trans. John Cottingham. Cambridge: Cambridge University Press, 1984.

Dessen, Alan C., and Leslie Thomson. *A Dictionary of Stage Directions in English Drama 1580–1642.* Cambridge: Cambridge University Press, 1999.

Dryden, John. *"Of Dramatic Poesy" and Other Critical Essays*. Ed. George Watson. London: Dent, 1962.

du Bois, Page. *Torture and Truth*. New York: Routledge, 1991.

Du Laurens, André. *A Discourse of the Preservation of the Sight: of Melancholike Diseases; of Rheumes, and of Old Age*. (1599) Trans. Richard Surphlet. Oxford: Oxford University Press, 1938.

Edelstein, Ludwig. *Ancient Medicine*. Baltimore: Johns Hopkins University Press, 1967.

Edmond, Mary. *Hilliard and Oliver: The Lives and Works of Two Great Miniaturists*. London: Hale, 1983.

Edwards, Anthony S. G., ed. *Skelton: The Critical Heritage*. London: Routledge and Kegan Paul, 1981.

Eliot, T. S. "Hamlet and His Problems." *The Sacred Wood*, 95–103. London: Methuen, 1920.

Elkins, James. *Pictures of the Body: Pain and Metamorphosis*. Stanford: Stanford University Press, 1999.

Empedocles. *Empedocles: The Extant Fragments*. Trans. M. R. Wright. New Haven: Yale University Press, 1981.

Engle, Lars. "'Thrift Is Blessing': Exchange and Explanation in *The Merchant of Venice*." *Shakespeare Quarterly* 37 (1986): 20–37.

Erasmus. *Ten Colloquies*. New York: The Liberal Arts Press, 1957.

Esoteric. "The Noise of Depression." *Epistemological Despondency*. Aesthetic Death, 1994.

Evans, Bergen. *The Psychiatry of Robert Burton*. New York: Columbia University Press, 1944.

Ferrand, Jacques. *A Treatise on Lovesickness*. Trans. and ed. Donald A. Beecher and Massimo Ciavolella. Syracuse: Syracuse University Press, 1990.

Ficino, Marsilio. *Three Books on Life: A Critical Edition and Translation with Introduction and Notes*. Ed. Carol V. Kaske and John R. Clark. Tempe: Renaissance Society of America, 2002.

Fink, Zera S. "Jacques and the Malcontent Traveler." *Philological Quarterly* 14 (July 1935): 237–252.

Fish, Stanley. *How Milton Works*. Cambridge, Mass.: Harvard University Press, 2001.

———. *Self-Consuming Artifacts: The Experience of Seventeenth-Century Literature*. Berkeley: University of California Press, 1972.

Flatley, Jonathan. *Affective Mapping: Melancholia and the Politics of Modernism*. Cambridge, Mass.: Harvard University Press, 2008.

Fox, Dominic. *Cold World: The Aesthetics of Dejection and the Politics of Militant Despondency*. London: Zero Books, 2009.

Fox, Ruth A. *The Tangled Chain: The Structure of Disorder in the Anatomy of Melancholy*. Berkeley: University of California Press, 1976.

Foucault, Michel. *Language, Counter-Memory, Practice: Selected Essays and Interviews*. Ed. Donald Bouchard. Ithaca: Cornell University Press, 1977.

Freccero, Carla. *Queer / Early / Modern*. Durham: Duke University Press, 2006.

Freeman, Elizabeth. "Packing History, Count(er)ing Generations." *New Literary History* 31 (2000): 727–744.

Freud, Sigmund. "A Child Is Being Beaten." *Collected Papers*. Vol. 2. New York: Basic Books, 1959.

———. *The Complete Letters of Sigmund Freud to Wilhelm Fleiss: 1887–1904*. Ed. Jeffrey Moussaieff Masson. Cambridge, Mass.: Harvard University Press, 1985.

———. "The Economic Problem of Masochism." *Standard Edition*. Trans. James Strachey. Vol. 19:157–173. London: The Hogarth Press, 1961.

———. *The Interpretation of Dreams*. Trans. Dr. A. A. Brill. New York: Gramercy Books, 1996.

———. "Mourning and Melancholia." *The Standard Edition of the Complete Psychological Works of Sigmund Freud*, trans. James Strachey, 14: 239–260. London: Hogarth Press, 1955.

———. "The Theme of the Three Caskets." *Collected Papers*. Vol. 4. New York: Basic Books. 1959.

———. *Three Essays on the Theory of Sexuality*. Trans. James Strachey. New York: Avon, 1962.

Fried, Michael. *Absorption and Theatricality: Painting and the Beholder in the Age of Diderot*. Berkeley: University of California Press, 1980.

Galen. *On the Affected Parts*. Trans. and ed. Rudolph E. Siegel. Basel: S Karger, 1976.

———. *Three Treatises on the Nature of Science*. Trans. Richard Walzer and Michael Frede. Indianapolis: Hackett Publishing Company, 1985.

Garcia-Ballester, Luis. *Galen and Galenism: Theory and Medical Practice from Antiquity to the European Renaissance*. Aldershot: Ashgate Variorum, 2002.

Gent, Lucy, and Nigel Llewellyn, eds. *Renaissance Bodies: The Human Figure in English Culture, 1540–1660*. London: Reaktion Books, 1990.

Gibbons, Brian, ed. *The Revenger's Tragedy*. New York: W. W. Norton, 1991.

Girard, René. "To Entrap the Wisest: A Reading of *The Merchant of Venice*." In *Literature and Society*, edited by Edward W. Said, 100–119. Baltimore: Johns Hopkins University Press, 1980.

Gottfried, Robert. *Doctors and Medicine in Medieval England: 1340–1530*. Princeton: Princeton University Press, 1986.

Grace, Wendy. "*Faux Amis*: Foucault and Deleuze on Sexuality and Desire." *Critical Inquiry* 36, no. 1 (Autumn 2009): 52–76.

Greenblatt, Stephen. *Learning to Curse: Essays in Early Modern Culture*. New York: Routledge, 1990.

———. *Shakespearean Negotiations: The Circulation of Social Energy in Renaissance England*. Berkeley: University of California Press, 1989.

———. *Will in the World.* New York: W. W. Norton, 2004.

Grose, Christopher. "Theatrum Libri: Burton's Anatomy of Melancholy and the Failure of Encyclopedic Form." In *Books and Readers in Early Modern England,* edited by Jennifer Andersen and Elizabeth Sauer. Philadelphia: University of Pennsylvania Press, 2001. 80–97.

Gross, Daniel. "Defending the Humanities with Charles Darwin's *The Expression of Emotions in Man and Animals* (1872)." *Critical Inquiry* 37, no. 1 (Autumn 2010): 34–60.

Guillory, John. "'To Please the Wiser Sort': Violence and Philosophy in *Hamlet.*" *Historicism, Psychoanalysis, and Early Modern Culture,* edited by Carla Mazzio and Douglas Trevor, 82–110. New York: Routledge, 2000.

Guilpin, Edward. *Skialetheia* (1598). Oxford: Shakespeare Association Facsimile, 1931.

Hacking, Ian. *The Social Construction of What?* Cambridge, Mass.: Harvard University Press, 2000.

Halperin, David. *How to Do the History of Homosexuality.* Chicago: University of Chicago Press, 2004.

Halpern, Richard. *Shakespeare among the Moderns.* Ithaca: Cornell University Press, 1997.

Hammill, Graham. *Sexuality and Form: Caravaggio, Marlowe, and Bacon.* Chicago: University of Chicago Press, 2000.

Hanson, Elizabeth. *Discovering the Subject in Renaissance England.* Cambridge: Cambridge University Press, 1998.

Hanssen, Beatrice. *Walter Benjamin's Other History: Of Stones, Animals, Human Beings, and Angels.* Berkeley: University of California Press, 2000.

Harbage, Alfred. "*Love's Labour's Lost* and the Early Shakespeare." *Philological Quarterly* 41 (1962): 18–36.

Harlow, Harry. *Learning to Love.* New York: Jason Aronson, 1974.

Harlow, H. F., J. McGaugh, and R. F. Thompson. *Psychology.* San Francisco: Albion, 1971.

Harman, Graham. "On Vicarious Causation." *Collapse Vol. II: Speculative Realism* (March 2007): 187–221.

———. *Prince of Networks: Bruno Latour and Metaphysics.* Melbourne: re.press, 2009.

———. *Towards Speculative Realism.* London: Zero Books, 2010.

Harris, Jonathan Gil. *Untimely Matter in the Time of Shakespeare.* Philadelphia: University of Pennsylvania Press, 2009.

Hegel, G. W. F. *On the Philosophy of Fine Art.* Vol. 2. Trans. F. P. B. Osmaston. Oxford: Oxford University Press, 1920.

Heidel, William. *Hippocratic Medicine.* New York: Columbia University Press, 1941.

Hejinian, Lyn. *The Fatalist.* New York: Omnidawn, 2003.

Herbert, Edward. *The Life of Edward, First Lord Herbert of Cherbury, Written by Himself.* Ed. J. M. Shuttleworth. London: Oxford University Press, 1976.

Here Is Always Somewhere Else: The Disappearance of Bas Jan Ader. Dir. Rene Daalder. Cult Epics, 2008.

Hillman, David. *Shakespeare's Entrails: Belief, Skepticism, and the Interior of the Body.* New York: Palgrave Macmillan, 2007.

Hillman, David, and Carla Mazzio, eds. *The Body in Parts: Fantasies of Corporeality in Early Modern Europe.* London: Routledge, 1997.

Hippocrates. *Pseudiepigraphic Writings.* Trans. Wesley D. Smith. Leiden: E. J. Brill, 1990.

Hirsh, J. E. "The 'To be, or not to be' Scene and the Conventions of Shakespeare's Drama." *Modern Language Quarterly* 42 (1981): 115–136.

Hobbes, Thomas. *Leviathan.* Ed. Richard Tuck. Cambridge: Cambridge University Press, 1996.

Hunter, G. K. "Notes on 'Asides' in Elizabethan Drama." *Notes and Queries* 1 (1997): 83–84.

Hutchison, Jane Campbell. *Albrecht Dürer: A Biography.* Princeton: Princeton University Press, 1990.

Inwood, Brad, and Lloyd P. Gerson, eds. *The Stoics Reader: Selected Writings and Testimonia.* Cambridge, Mass.: Hackett Publishing Co., 2008.

Jackson, Stanley. *Melancholia and Depression.* New Haven: Yale University Press, 1986.

James, William. *The Will to Believe and Other Essays in Popular Philosophy.* Cambridge, Mass.: Harvard University Press, 1979.

Jones, Ernest. *Hamlet and Oedipus.* London: W. W. Norton, 1949.

Jonson, Ben. *Every Man in His Humour: A Parallel-Text Edition of the 1601 Quarto and the 1616 Folio.* Ed. J. W. Lever Lincoln: University of Nebraska Press, 1971.

Jordan-Smith, Paul. *Bibliographia Burtoniana: A Study of Robert Burton's* The Anatomy of Melancholy, *with a Bibliography of Burton's Writings.* Stanford: Stanford University Press, 1931.

Kagan, Jerome. *What Is Emotion?: History, Measures, and Meanings.* New Haven: Yale University Press, 2007.

Kahn, Victoria. "Aesthetics as Critique: Tragedy and *Trauerspiel* in *Samson Agonistes.*" In *Reading Renaissance Ethics,* edited by Marshall Grossman, 104–129. New York: Routledge, 2007.

———. *Wayward Contracts: The Crisis of Political Obligation in England, 1640–1674.* Princeton: Princeton University Press, 2004.

Kelley, Mark, Michael Lieb, and John Shawcross, eds. *Milton and the Grounds of Contention.* Pittsburgh: Duquesne University Press, 2003.

Kitzes, Adam. *The Politics of Melancholy from Spenser to Milton.* London: Routledge, 2006.

Klibansky, Raymond, Erwin Panofsky, and Fritz Saxl. *Saturn and Melancholy: Studies in the History of Natural Philosophy, Religion and Art*. London: Nelson, 1964.

Knott, John R. *Discourses of Martyrdom in English Literature, 1563–1694*. Cambridge: Cambridge University Press, 1993.

Krafft-Ebbing, Richard. *Psychopathia Sexualis: A Medico-Forensic Study*. New York: Putnam, 1965.

Kristeva, Julia. *Black Sun: Depression and Melancholia*. New York: Columbia University Press, 1989.

Lacan, Jacques. *Seminar Book X: Anxiety 1962-63*. Trans. Cormac Gallagher. London: Antony Rowe, 2004.

Langley, Eric. *Narcissism and Suicide in Shakespeare and His Contemporaries*. Oxford: Oxford University Press, 2009.

Laplanche, J., and J-B Pontalis. *The Language of Psychoanalysis*. Trans. Donald Nicholson-Smith. New York: W. W. Norton, 1973.

Laplanche, Jean. *Seduction, Translation and Drives*. Trans. Martin Stanton. London: Psychoanalytic Forum, Institute of Contemporary Arts, 1992.

Laqueur, Thomas. *Making Sex: Body and Gender from the Greeks to Freud*. Cambridge, Mass.: Harvard University Press, 1992.

Leader, Darian. *The New Black: Mourning, Melancholia and Depression*. New York: Graywolf, 2009.

Lemnius, Levinus. *The Touchstone of Complexions: Generallye Appliable, Expedient and Profitable for all Such as be desirous and carefull of theyr bodyly health. Englished by Thomas Newton*. London: Thomas Marsh, 1581.

Lenzer, Gertrud. "On Masochism: A Contribution to the History of a Phantasy and Its Theory." *Signs: A Journal of Women in Culture and Society* 1 (1975): 277–324.

Lepenies, Wolf. *Melancholy and Society*. Cambridge, Mass.: Harvard University Press, 1992.

Levao, Ronald. *Renaissance Minds and Their Fictions*. Berkeley: University of California Press, 1985.

Lewis, Cynthia. "Antonio and Alienation in 'The Merchant of Venice.'" *South Atlantic Review* 48.4 (1983): 19–31.

Leys, Ruth. "How Did Fear Become a Scientific Object and What Kind of Object Is It?" *Representations* 110 (Spring 2010): 66–104.

Liddell, Henry George, and Robert Scott, eds. *A Greek-English Lexicon*. Oxford: Clarendon Press, 1996.

Loewenstein, David. "From Politics to Faith in the Great Poems?" In *Visionary Milton: Essays on Prophecy and Violence*, edited by Peter E. Medine, John Shawcross, and David V. Urban, 269–287. Pittsburgh: Duquesne University Press, 2010.

Lomazzo, Paolo Giovanni. *A Tracte Containing the Artes of Curious Painting* (1598). New York: Da Capo Press, 1969.

Londré, Felicia Hardison, ed. *Love's Labour's Lost: Critical Essays.* New York: Garland, 1997.

Lund, Mary Ann. *Melancholy, Medicine, and Religion in Early Modern England: Reading the Anatomy of Melancholy.* Cambridge: Cambridge University Press, 2010.

Lupton, Julia. "Creature Caliban." *Shakespeare Quarterly* 51 (2000): 1–24.

———. "Samson Dagonistes." In *Citizen-Saints: Shakespeare and Political Theology,* 181–205. Chicago: University of Chicago Press, 2005.

———. *Thinking with Shakespeare: Essays on Politics and Life.* Chicago: University of Chicago Press, 2011.

Lupton, Julia, and Kenneth Reinhard. *After Oedipus: Shakespeare in Psychoanalysis.* Ithaca: Cornell University Press, 1993.

Lyons, Bridget Geller. *Voices of Melancholy: Studies in Literary Treatments of Melancholy in Renaissance England.* New York: W. W. Norton, 1971.

Lytton, Edward Bulwer. *The Last Days of Pompeii.* Paris: Baudry's European Library, 1839.

Marston, John. "The Metamorphosis of Pigmalion's Image, and Certain Satyres." In *Works.* Vol. 3, 200–237. London: John Russell Smith, 1856.

Masciandaro, Nicola. "Anti-Cosmosis: Black Mahapralaya." *Hideous Gnosis: Black Metal Theory Symposium I.* New York: CreateSpace, 2010.

Maus, Katherine Eisaman. *Inwardness and Theater in the English Renaissance.* Chicago: University of Chicago Press, 1995.

Mazzio, Carla. *The Inarticulate Renaissance: Language Trouble in an Age of Eloquence.* Philadelphia: University of Pennsylvania Press, 2009.

McNally, Richard. *What Is Mental Illness?* Cambridge, Mass.: Harvard University Press, 2011.

Meillassoux, Quentin. *After Finitude: An Essay on the Necessity of Contingency.* Trans. Ray Brassier. London: Continuum, 2008.

Melancholia. Dir. Lars von Trier. Perf. Kirsten Dunst, Charlotte Gainsbourg, and Alexander Skarsgard. Zentropa Entertainments, 2011.

Merleau-Ponty, M. *Phenomenology of Perception.* Trans. Colin Smith. London: Routledge, 1962.

Milton, John. *The Complete Poetry and Essential Prose.* Ed. William Kerrigan, John Rumrich, and Stephen M. Fallon. New York: Modern Library, 2007.

———. *The Complete Works of John Milton. Vol. II: The 1671 Poems; Paradise Regain'd and Samson Agonistes.* Ed. Laura Lunger Knoppers. Oxford: Oxford University Press, 2008.

Mohamed, Feisal G. "Confronting Religious Violence: Milton's 'Samson Agonistes.'" *PMLA* 120, no. 2 (2005): 327–340.

———. *Milton and the Post-Secular Present: Ethics, Politics, Terrorism.* Stanford: Stanford University Press, 2011.

Moncure-Sime, A. H. *Shakespeare: His Music and Song.* London: K. Paul, Trench, Trübner, 1915.

Morton, Timothy. *Ecology without Nature: Rethinking Environmental Aesthetics.* Cambridge, Mass.: Harvard University Press, 2007.

Müller, Christopher. *Bas Jan Ader.* Cologne: Verlag Der Buchhandlung Walther Konig, Kunstverein Braunschweig, Bonner and Kunstverein Munchen, 2000.

Noe, Alva. *Out of Our Heads: Why You are Not Your Brain, and Other Lessons from the Biology of Consciousness.* New York: Hill and Wang, 2010.

O'Connell, Michael. *Robert Burton.* Boston: Twayne Publishers, 1986.

Panofsky, Erwin. *The Life and Art of Albrecht Dürer.* Princeton: Princeton University Press, 1943.

Parker, David. *The Commonplace Book in Tudor London.* New York: University Press of America, 1998.

Paster, Gail Kern. *Humoring the Body: Emotions and the Shakespearean Stage.* Chicago: University of Chicago Press, 2004.

Pater, Walter. *Appreciations: With an Essay on Style.* London: Macmillan, 1898.

Patterson, Steve. "The Bankruptcy of Homoerotic Amity in Shakespeare's *Merchant of Venice.*" *Shakespeare Quarterly* 50 (1999): 9–32.

Plato. *Phaedrus.* Trans. and ed. James H. Nichols Jr. Ithaca: Cornell University Press, 1998.

Pointon, Marcia. *Hanging the Head: Portraiture and Social Formation in Eighteenth-Century England.* New Haven: Yale University Press, 1993.

Polanyi, Michael. *The Tacit Dimension.* Chicago: University of Chicago Press, 2009.

Prescott, Anne Lake. "Pierre de la Primaudaye's French Academy: Growing Encyclopedic." In *The Renaissance Computer: Knowledge Technology in the First Age of Print,* edited by Neil Rhodes and Jonathan Sawday, 157–167. London: Routledge, 2000.

Pye, Christopher. *The Vanishing: Shakespeare, the Subject, and Early Modern Culture.* Durham: Duke University Press, 2000.

Radden, Jennifer. *The Nature of Melancholy: From Aristotle to Kristeva.* Oxford: Oxford University Press, 2002.

Rambuss, Richard. *Closet Devotions.* Durham: Duke University Press, 1998.

Reik, Theodor. *Masochism in Sex and Society.* Trans. Margaret H. Beigel and Gertrud M. Kurth. New York: Grove Press, 1962.

Renaker, David. "Robert Burton and Ramist Method." *Renaissance Quarterly* 24, no. 2 (Summer 1971): 210–220.

Robin, P. Ansell. *The Old Physiology in English Literature.* London: J. M. Dent and Sons, 1911.

Rosen, Charles. "The Ruins of Walter Benjamin." In *On Walter Benjamin: Critical Essays and Recollections,* edited by Gary Smith. Cambridge, Mass.: The MIT Press, 1991. 129–176.

Sainsbury, R. M. "Concepts without Boundaries." In *Vagueness: A Reader*, edited by Rosanna Keefe and Peter Smith. Cambridge, Mass.: The MIT Press, 1996. 251–265.

Santner, Eric. *On Creaturely Life: Rilke, Benjamin, Sebald*. Chicago: University of Chicago Press, 2006.

Sarton, George. *Galen of Pergamon*. Kansas City: University of Kansas Press, 1954.

Sartre, Jean-Paul. *Sketch for a Theory of the Emotions*. Trans. Philip Mairet. London: Routledge, 2003.

Sawday, Jonathan. *The Body Emblazoned: Dissection and the Human Body in Renaissance Culture*. London: Routledge, 1995.

———. "Shapeless Eloquence: Burton's Anatomy of Knowledge." In *English Renaissance Prose: History, Language and Politics*, edited by Neil Rhodes. Tempe: Arizona State University Press, 1997. 173–202.

Schiesari, Juliana. *The Gendering of Melancholia: Feminism, Psychoanalysis, and the Symbolics of Loss in Renaissance Literature*. Ithaca: Cornell University Press, 1992.

Schlegel, Augustus Wilhelm von. *Lectures on Dramatic Art and Literature*. Ed. Reverend A. J. W. Morrison. London: George Bell & Sons, 1879.

Schleiner, Winfried. *Melancholy, Genius and Utopia in the Renaissance*. Wiesbaden: Harrasowitz, 1991.

Schmelzer, Mary Murphy. *'Tis All One: "The Anatomy of Melancholy" as Belated Copious Discourse*. New York: Peter Lang, 1999.

Schmitt, Carl. *Political Theology*. Trans. George Schwab. Chicago: University of Chicago Press, 1985.

Schoenfeldt, Michael C. *Bodies and Selves in Early Modern England: Physiology and Inwardness in Spenser, Shakespeare, Herbert and Milton*. Cambridge: Cambridge University Press, 1999.

Scott, Jonathan. *Salvator Rosa: His Life and Times*. New Haven: Yale University Press, 1996.

Searle, John. *Making the Social World*. Oxford: Oxford University Press, 2010.

Sedgwick, Eve Kosofsky. "Introduction: Axiomatic." In *The Epistemology of the Closet*, 1–67. Berkeley: University of California Press, 1990.

Seeling, Sharon Cadman. "Burton: *The Anatomy of Melancholy*: 'I Have Overshot My Selfe.'" *Generating Texts: The Progeny of Seventeenth-Century Prose*. University of Virginia Press, 1996.

Shakespeare, William. *Love's Labour's Lost*. Ed. G. R. Hibbard. Oxford: Clarendon Press, 1990.

———. *The Merchant of Venice*. Ed. M. M. Mahood. Cambridge: Cambridge University Press, 1987.

———. *The Merchant of Venice*. Ed. John Russell Brown. London: The Arden Shakespeare, 1955.

———. *The Norton Shakespeare*. Ed. Stephen Greenblatt. New York: W. W. Norton, 1997.

———. *The Tragedy of Hamlet, Prince of Denmark*. Ed. Susanne Wofford. New York: St. Martin's, 1994.

———. *The Tragedy of Hamlet, Prince of Denmark*. Ed. Harold Jenkins. London: Methuen, 1982.

Sharpe, Ellie. *Collected Papers on Psychoanalysis*. Ed. Marjorie Brierly. London: Hogarth Press, 1950.

Shohet, Lauren. "YouTube, Use, and the Idea of the Archive." In *Shakespeare Studies*, 68–76. Volume XXXVIII. Madison: Fairleigh Dickinson University Press, 2010.

Shuger, Deborah. *Habits of Thought in the English Renaissance: Religion, Politics, and the Dominant Culture*. Toronto: University of Toronto Press, 1997.

Silliman, Ron. "For L=A=N=G=U=A=G=E." In *The* L=A=N=G=U=A=G=E Book, edited by Bruce Andrews and Charles Bernstein. Carbondale: Southern Illinois University Press, 1984.

Silverman, Kaja. "Masochism and Male Subjectivity." In *Male Subjectivity at the Margins*, 185–214. London: Routledge, 1992.

Siraisi, Nancy. *Medieval and Early Renaissance Medicine*. Chicago: University of Chicago Press, 1990.

Slights, William. *The Heart in the Age of Shakespeare*. Cambridge: Cambridge University Press, 2008.

Smirnoff, Victor. "The Masochistic Contract." In *Essential Papers on Masochism*, edited by Margaret Hanley, 62–75. New York: New York University Press, 1995.

Sontag, Susan. *Under the Sign of Saturn*. New York: Vintage Books, 1972.

Speak, Gill. "An Odd Kind of Melancholy: Reflections on the Glass Delusion in Europe (1440–1680)." *History of Psychiatry* 1, no. 2 (June 1990): 191–206.

Spinoza, Benedictus de. *A Spinoza Reader: The Ethics and Other Works*. Ed. Edwin Curley. Princeton: Princeton University Press, 1994.

Stachniewski, John. *The Persecutory Imagination: English Puritanism and the Literature of Religious Despair*. Oxford: Clarendon Press, 1991.

Strauss, Walter. *The Complete Engravings, Etchings & Drypoints of Albrecht Dürer*. New York: Dover Publications, 1972.

Striborg. "Digging a Ditch To Die In." *Embittered Darkness / Isle De Morts*. Displeased Records / Southern Lord Recordings, 2006.

Strong, Roy. *Artists of the Tudor Court: The Portrait Miniature Rediscovered 1520–1620*. London: The Victoria & Albert Museum, 1983.

———. *The English Renaissance Miniature*. London: Thames and Hudson, 1983.

Sugimura, N. K. *'Matter of Glorious Trial': Spiritual and Material Substance in Paradise Lost*. New Haven: Yale University Press, 2009.

Taylor, Sue. *Hans Bellmer: The Anatomy of Anxiety*. Cambridge, Mass.: The MIT Press, 2000.

Teskey, Gordon. *Delirious Milton: The Fate of the Poet in Modernity*. Cambridge, Mass.: Harvard University Press, 2006.

Tofte, Robert. *Alba: The Month's Minde of a Melancholy Lover, Divided into Three Parts. By R. T. Gentleman*. London: Printed by Felix Kingston, for Matthew Lownes, 1598.

Trevor, Douglas. *The Poetics of Melancholy in Early Modern England*. Cambridge: Cambridge University Press, 2004.

Turner, Henry S. "The Problem of the More-Than-One: Friendship, Calculation, and Political Association in *The Merchant of Venice*." *Shakespeare Quarterly* 57, no. 4 (2006): 413–442.

Verwoert, Jan. *Bas Jan Ader: In Search of the Miraculous*. London: Afterall Books, 2006.

Ware, Owen. "Love Speech." *Critical Inquiry* 34, no. 3 (Spring 2008): 491–509.

Weitz, Morris. "Hamlet: Philosophy, The Intruder." *Shakespeare, Philosophy, and Literature*, edited by Margaret Collins Weitz, 17–33. New York: Lang, 1995.

Wells, Stanley, and Gary Taylor, eds. *William Shakespeare: A Textual Companion*. Oxford: Clarendon Press, 1987.

Wendorf, Richard. *The Elements of Life: Biography and Portrait-Painting in Stuart and Georgian England*. Oxford: Oxford University Press, 1992.

Wheeler, Thomas, ed. *The Merchant of Venice: Critical Essays*. New York: Garland, 1991.

Williams, Raymond. "Structures of Feeling." In *Marxism and Literature*, 128–136. Oxford: Oxford University Press, 1977.

Wilson, Scott, ed. *Melancology: Black Metal Theory and Ecology*. London: Zero Books, 2013.

Wilson, John Dover. *What Happens in Hamlet*. Cambridge: Cambridge University Press, 1995.

Winterstein, Alfred. *Dürer's 'Melancholie' im Lichte der Psychoanalyse*. Leipzig: Internationaler Psychoanalytischer Verlag, 1929.

Wittgenstein, Ludwig. *Philosophical Investigations*. Trans. Elizabeth Anscombe. Oxford: Basil Blackwell, 1963.

Žižek, Slavoj. *For They Know Not What They Do: Enjoyment as a Political Factor*. London: Verso, 1991.

———. *Looking Awry: An Introduction to Jacques Lacan through Popular Culture*. Cambridge, Mass.: The MIT Press, 1992.

———. "Is There a Cause of the Subject?" In *Supposing the Subject*, edited by Joan Copjec. London: Verso, 1994. 84–106.

Zupančič, Alenka. "When Surplus Enjoyment Meets Surplus Value." In *Jacques Lacan and the Other Side of Psychoanalysis: Reflections on 'Seminar XVII,'* edited by Justin Clemens and Russell Grigg, 155–178. Durham: Duke University Press, 2006.

Index

www.ingramcontent.com/pod-product-compliance
Ingram Content Group UK Ltd.
Pitfield, Milton Keynes, MK11 3LW, UK
UKHW020644230225
455416UK00011B/62

9 780823 251278